Springer Texts in Business and Economics

More information about this series at
http://www.springer.com/series/10099

Martin Kolmar · Magnus Hoffmann

Workbook for Principles of Microeconomics

Martin Kolmar
School of Economics
University of St. Gallen
St. Gallen, Switzerland

Magnus Hoffmann
School of Economics
University of St. Gallen
St. Gallen, Switzerland

ISSN 2192-4333 ISSN 2192-4341 (electronic)
Springer Texts in Business and Economics
ISBN 978-3-319-62661-1 ISBN 978-3-319-62662-8 (eBook)
https://doi.org/10.1007/978-3-319-62662-8

Library of Congress Control Number: 2018930104

Printed on acid-free paper

This Springer imprint is published by Springer Nature
The registered company is Springer International Publishing AG
The registered company address is: Gewerbestrasse 11, 6330 Cham, Switzerland

Acknowledgements

If you want to learn the piano, you have to sit down and start practicing. This is tedious in the beginning and will not sound particularly pretty but things will improve over time. The same applies to economics (and other scientific theories as well). If you want to use them to better understand and analyze certain aspects of economic and social reality, you have to make them your own, understand their internal "mechanics" and work with them. Reading textbooks or listening to lectures is only a poor substitute. Limiting yourself to it would be like wanting to learn how to play piano by just listening to a piano player and studying piano scores.

This is why we have collected a series of problems and exercises that are intended to help you to adopt step by step the theories introduced and discussed in the textbook "Principles of Microeconomics: An Integrative Approach". You will find a chapter with different types of problems and sample solutions that corresponds to a chapter in the main book. We distinguish between three different types of exercises that focus on the development of specific and complementary skills and competencies.

The first type is true or false exercises; statements that can either be true or false. At the end of each section you will find the solutions along with short explanations, as well as links to the textbook.

The second type of problems have the character of short case studies or word problems, to answer which you are will be required to develop a more complex train of thoughts. Problems like these do not have one and only one correct solution but can usually be approached from different directions. Nevertheless, this book offers you sample solutions at the end of this section that illustrate *one* possible approach. Over the years during which we have developed the problems and used them in class we have also been able to identify typical lines of faulty reasoning. We will look into these and explain how they can be avoided.

Finally, you will find multiple-choice questions to answer in which you will have to identify one correct answer from a choice of given answers. Please note that all references to chapters are to those in the textbook "Principles of Microeconomics: An Integrative Approach" by Martin Kolmar, unless otherwise specified.

The teaching material collected in this book has grown over many years and is a result of efforts of a great many people, first of all the students who worked with them. We are most thankful for their innumerable suggestions that helped us improving the set of problems. Further, we thank Dario Fauceglia, Jürg Furrer, Carolin Güssow, Katharina Hofer, Alfonso Sousa-Poza, and Fred Henneberger for pointing out many errors, inconsistencies, and ways to improve the problems included in this book. Last but not least we thank our student assistants Corinne Knöpfel, Jan Riss, and Jan Serwart without whom the book would not be as it is.

Science that aims at both better understanding the reality and practical application of theory is similar to jazz. A good economist is like a good pianist: both have to master their instruments to be able to improvise. You know you're there when it starts to swing. We hope that this book will help you on your way to reaching this goal.

St. Gallen, in July 2017 Martin Kolmar and Magnus Hoffmann

Contents

1 First Principles

1.1 True or False

1.1.1 Statements

1.1.1.1 Block 1

1. The more aspects of reality are taken into consideration, the more useful an economic model is.
2. According to Karl Popper, a basic requisite for the quality of scientific theories is that one can refute them.
3. Concerning logical statements: one can derive a false hypothesis from false assumptions.
4. Modern microeconomics is macro-founded.

1.1.1.2 Block 2

1. Economics, as a positive science, tries to explain why social phenomena work the way they work. Economics, as a negative science, tries to explain why social phenomena do not work the way they work.
2. If an economist tries to determine how a country should increase taxes in the best possible way, she or he is practicing normative science.
3. If an assumption in a scientific theory is incorrect, the theory must be discarded because it cannot contribute to the understanding of reality.
4. In economics, one investigates the interplay of human behavior on the individual level.

1.1.1.3 Block 3

1. With regards to economics, positive science answers the question as to how humans should cope with the phenomena of scarcity.

M. Kolmar, M. Hoffmann, *Workbook for Principles of Microeconomics*,
Springer Texts in Business and Economics, https://doi.org/10.1007/978-3-319-62662-8_1

2. Quantities of goods that are not on the production-possibility frontier, cannot be produced.
3. Modern macroeconomics is not "micro-founded" because it concentrates on economic aggregates.
4. Opportunity costs are costs of the past that cannot be influenced any longer.

1.1.1.4 Block 4

1. According to the theory of critical rationalism, the monopoly theory is not a scientifically sound theory, because the assumption of profit maximization has been falsified.
2. According to the theory of critical rationalism, scientific theories can be verified, but cannot be falsified.
3. According to the theory of critical rationalism, a good theory can be falsified in principle, but has not been falsified yet.
4. The implication of applying Ockham's razor to scientific theories is that a theory with fewer assumptions is preferable to one with more assumptions if both theories lead to identical hypotheses.

1.1.2 Solutions

1.1.2.1 Sample Solutions for Block 1

1. **False**. An important criterion for good models is simplicity or frugality. The idea is often referred to as "Ockham's razor" which states that, among competing models, the one with the fewest assumptions should be selected. Ockham's razor necessarily implies that the assumptions of a model should not be realistic in the naïve sense that the assumptions shall fit reality. See Chapter 1.2.3.
2. **True**. According to Karl Popper, scientific theories can never be verified but can, in principle, be falsified by bringing in empirical evidence that is in conflict with the hypotheses of the theory. See Chapter 1.2.6.
3. **True**. See examples in Chapter 1.2.2.
4. **False**. Modern macroeconomics is micro-founded, the converse argument does not hold. See Chapter 1.1.

1.1.2.2 Sample Solutions for Block 2

1. **False**. The goal of positive theories is explaining phenomena. Normative theories, on the other hand, try to determine what people should do in which situation. Thus, they are based on a judgement. See Chapter 1.2.7.
2. **True**. See the sample solution to Block 2, Statement 1.
3. **False**. Assumptions are necessarily simplifications. The important thing is finding the correct balance between reasonable simplification of assumptions and the underlying causal mechanisms on the one hand, and the content explained by the derived hypotheses on the other hand. See Chapter 1.2.4.

4. **False**. On the individual level, one investigates individual humans' behavior. Humans' behavioral interplay is investigated on the interaction level. See Chapter 1.1.

1.1.2.3 Sample Solutions for Block 3

1. **False**. With regards to economics, positive science answers the question as to how humans cope (without judgement) with the phenomena of scarcity. See Chapter 1.2.7.
2. **False**. The production-possibility frontier indicates the maximum quantity that can be produced. Every quantity below the frontier can be produced as well. See Chapter 1.2.5.
3. **False**. The micro-foundation of macroeconomics is a research program that tries to explain the regularities on the aggregate level, like relationships between inflation and unemployment, through individuals' behavior and interactions. The current macroeconomics mainstream is, in this sense, largely micro-founded. See Chapter 1.1.
4. **False**. Opportunity costs are costs that result from, for example, forgoing an alternative use of capital or time (such as the salary that a student forgoes, because he is not working). See Chapter 1.1.

1.1.2.4 Sample Solutions for Block 4

1. **False**. Falsification is disproving hypotheses by confronting them with empirical evidence that conflict with the hypotheses. One can only falsify hypotheses, not assumptions. See Chapter 1.2.6.
2. **False**. According to the theory of critical rationalism, scientific theories can never be irrevocably proven, but they can in principle be falsified. See Chapter 1.2.6.
3. **True**. Theories should be formulated such that their hypotheses are falsifiable. Good theories are those that have a large empirical content, but have not been falsified so far. See Chapter 1.2.6.
4. **True**. This is true by definition. See Chapter 1.2.3 and the sample solution to Block 1, Statement 1.

2 Gains From Trade

2.1 True or False

2.1.1 Statements

2.1.1.1 Block 1

There are two individuals, A and B, who can produce two goods, 1 and 2. The production-possibility frontiers of both individuals are $x_1^A = a - b \cdot x_2^A$ and $x_1^B = c - d \cdot x_2^B$, in which a, b, c and d are strictly larger than zero.

1. If $b > d$, then A has a comparative advantage in the production of good 1.
2. If $a > c$, then A has an absolute advantage in the production of both goods.
3. If $a = c$, then no individual has a comparative advantage.
4. If $a = 100$ and $b = 2$, then A can produce 50 units of the second good at maximum.

2.1.1.2 Block 2

1. A situation in which there is no trade between countries is defined as "autarky."
2. The theory of comparative advantage is only valid for linear production-possibility frontiers.
3. If a country has a comparative disadvantage in the production of a good, it should not trade this good with other countries.
4. All countries always benefit from specialization and trade.

2.1.1.3 Block 3

Charlotte and Phil are both bakers. Charlotte can either bake 20 cakes, 15 pizzas or any linear combination of the two in one day. Phil can either bake 10 cakes, 5 pizzas or any linear combination of the two.

1. Charlotte has a comparative advantage in baking pizza.
2. Charlotte has an absolute advantage in baking pizza.

M. Kolmar, M. Hoffmann, *Workbook for Principles of Microeconomics*,
Springer Texts in Business and Economics, https://doi.org/10.1007/978-3-319-62662-8_2

3. Phil's opportunity costs for a pizza are equivalent to two cakes.
4. Charlotte's opportunity costs for cake are lower than Phil's.

2.1.1.4 Block 4

1. Assume linear production-possibility frontiers. If two individuals have identical opportunity costs, then neither individual has a comparative advantage.
2. When compared with autarky, two individuals are never worse off if they specialize according to their comparative advantage and subsequently trade.
3. The theory of comparative advantage describes and explains the international trade of goods.
4. The sequence of the integration in a trade community (using a sequential procedure) is irrelevant for the trading partners' assessment of the advantageousness of the community.

2.1.1.5 Block 5

1. Individuals A and B can both produce either roses or computers. If they can become better off by trading, then one of the individuals will consign to only producing roses, while the other will consign to solely producing computers.
2. A comparison of opportunity costs allows one to identify potential absolute advantages.
3. An individual can have a comparative advantage concerning one good and an absolute advantage concerning a different good.
4. Two individuals with identical linear production-possibility frontiers can be better off by trading with each other.

2.1.2 Solutions

2.1.2.1 Sample Solutions for Block 1

Opportunity costs of good i in units of good j for individual k are $OC_{ij}^k = \left|\frac{dx_j^k}{dx_i^k}\right|$, where $i, j \in \{1,2\}$ and $k \in \{A, B\}$. Individual A's opportunity costs are then $OC_{12}^A = \frac{1}{b}$ and $OC_{21}^A = b$. Individual B's opportunity costs are then $OC_{12}^B = \frac{1}{d}$ and $OC_{21}^B = d$. See Chapter 2.2.

1. **True**. If $b > d$, then A's opportunity costs for good 1 are lower than B's and, thus, A has a comparative advantage in the production of good 1.
2. **False**. One cannot determine that A has an absolute advantage in the production of the second good solely based on $a > c$.
3. **False**. One cannot derive that conclusion from $a = c$. In order for that to be the case, $b = d$ must hold as well.
4. **True**. Individual A can produce $x_1^A = 100 - 2 \cdot x_2^A$ units of good 1. If A produces zero units of the first good, this would mean $0 = 100 - 2 \cdot x_2^A$ and thus $x_2^A = 50$.

2.1.2.2 Sample Solutions for Block 2

1. **True**. This is true by definition. See Chapter 2.1.
2. **False**. See the discussion about strictly concave and strictly convex production-possibility frontiers in Chapter 2.3.
3. **False**. That is exactly where an individual is able to be better off through trade. Because a comparative disadvantage in the production of one good always implies a comparative advantage in the production of another good. The individual can then specialize in the production of the good that he or she has a comparative advantage in and become better off due to trade. See Chapter 2.1.
4. **False**. The sequence of integration plays a role as well. Additionally, countries with identical opportunity costs will not have any gains from trade from trading with each other. See Chapter 2.3.

2.1.2.3 Sample Solutions for Block 3

Opportunity costs of good i in units of good j for individual k are $OC_{ij}^k = \left|\frac{dx_j^k}{dx_i^k}\right|$, where $i, j \in \{C, P\}$ and $k \in \{Ch, Ph\}$. Charlotte's opportunity costs are then $OC_{CP}^{Ch} = \frac{15}{20} = \frac{3}{4}$ and $OC_{PC}^{Ch} = \frac{20}{15} = \frac{4}{3}$. Phil's opportunity costs are then $OC_{CP}^{Ph} = \frac{5}{10} = \frac{1}{2}$ and $OC_{PC}^{Ph} = \frac{10}{5} = 2$. See Chapter 2.2.

1. **True**. $OC_{PC}^{Ch} = \frac{4}{3} < 2 = OC_{PC}^{Ph}$. Thus, Charlotte has a comparative advantage when baking pizza.
2. **True**. Charlotte can bake 15 pizzas, while Phil can only bake 5. Therefore, Charlotte has an absolute advantage.
3. **True**. $OC_{PC}^{Ph} = 2$.
4. **False**. $OC_{CP}^{Ch} = \frac{3}{4} > \frac{1}{2} = OC_{CP}^{Ph}$.

2.1.2.4 Sample Solutions for Block 4

1. **True**. Given identical opportunity costs, both individuals have to curb the production of one good by the same amount in order to produce one more unit of the other good. Consequently, no individual has a comparative advantage. See Chapter 2.1.
2. **True**. By specializing in one's comparative advantage, one is able to produce a surplus, which makes at least one of the individuals better off. The consumption under autarky can always be guaranteed. See Chapter 2.3.
3. **True**. The concept of comparative advantage cannot only be applied to individuals, but also to countries. See Chapter 2.1.
4. **False**. It is very relevant. See the detailed discussion about sequential integration in Chapter 2.3.

2.1.2.5 Sample Solutions for Block 5

1. **False**. If two individuals can become better off by trading, then there is a comparative advantage. The individuals will specialize according to their comparative advantage; however, whether they will completely specialize or not depends on their consumption preferences. See Chapter 2.2.
2. **False**. See Chapter 2.2.
3. **True**. Consider the example in Chapter 2.2, where Ann has both an absolute advantage in the production of tomatoes as well as a comparative advantage in the production of pears.
4. **False**. Because neither individual has a comparative advantage, neither of them can be better off through specialization and trade. See Chapter 2.1.

2.2 Open Questions

2.2.1 Problems

2.2.1.1 Exercise 1

Explain the theory of comparative advantage. Point out the theory's importance for economics, business administration, and law.

2.2.1.2 Exercise 2

There are two goods, 1 and 2, and two countries, A and B. Both goods are homogeneous and can be produced by both countries using labor as the only input, with each worker supplying 1 unit of labor. Each worker in A can produce 10 units of good 1, 10 units of good 2, or any linear combination of the two. In B, each worker can produce α units of good 1, 9 units of good 2, or any linear combination of the two. There are 100 workers in each country. The gains from trade are distributed among all workers in a country in a manner that makes everybody better off.

1. Determine and draw both countries' production-possibility frontiers for $\alpha = 8$.
2. Determine each country's opportunity costs of producing goods 1 and 2 for any given $\alpha > 0$.
3. Determine each country's comparative advantage depending on α.
4. Individuals in both countries always consume goods 1 and 2 in equal quantities. Determine the optimal production and consumption plans for $\alpha = 9$.
5. Assume that $\alpha = 10$. Show that both countries are better off, in comparison to autarky, when completely specializing in producing the good for which they have a comparative advantage.

2.2.1.3 Exercise 3

There are three countries, A, B and C, and each of them can produce two goods, 1 and 2. The production-possibility frontiers are given as:

$$x_1^A = 1 - x_2^A,$$
$$x_1^B = 1 - \tfrac{1}{2}x_2^B,$$
$$x_1^C = 1 - \tfrac{1}{4}x_2^C.$$

Each county is inhabited by individuals who always consume both goods in equal amounts. Potential gains from trade are distributed equally among the inhabitants of a country.

1. Determine the countries' production and consumption plans in autarky.
2. Assume countries A and C establish a free-trade agreement, such that a good produced in one of the two countries can be sold in both countries (without additional shipping costs). Determine the optimal production and consumption plans in these two countries if:
   a) gains from trade are split equally between the countries, with consumption under autarky (see Question 1) serving as a reference point.
   b) country C reaps all the benefits alone.
3. Assume countries A, B and C establish a free-trade agreement. Determine the optimal production and consumption plans of the countries if
   a) gains from trade are split equally, with consumption under autarky (see Question 1) serving as a reference point.
   b) gains from trade are split equally, with consumption under the first AC-agreement (see Question 2a)) serving as a reference point.
   c) gains from trade are split equally, with consumption under the second AC-agreement (see Question 2b)) serving as a reference point.
4. What are the implications for trade policy?

2.2.2 Solutions

2.2.2.1 Solutions to Exercise 1

In general, each individual has a comparative advantage in the production of one good, irrespective of whether she has an absolute advantage in the production of a good or not. An individual has a comparative advantage in the production of a given good if she can produce the good at lower opportunity costs, measured in units of the other good, than the other individuals. According to comparative advantages and trade, specialization has the potential to make all individuals better off. Therefore, the concept of comparative advantage is the basis for all further

Table 2.1 Exercise 2.2. Opportunity costs

	Country A	Country B
OC_{12}^{k}	$OC_{12}^{A} = 1$	$OC_{12}^{B} = \frac{9}{\alpha}$
OC_{21}^{k}	$OC_{21}^{A} = 1$	$OC_{21}^{B} = \frac{\alpha}{9}$

contemplation of the organization of economic activities, because one has to ask how economic activities have to be organized to allow specialization and trade.

Significance for Economics
Economics tries to understand how human beings organize (positive) or should organize (normative) economic activities in order to cope with the phenomenon of scarcity. The theory of comparative advantage explains why the organization of economic activities is at the core of economics and helps with the development of hypotheses about the purpose of organizations.

Significance for Business Administration
Business administration analyzes the unit of a firm or a corporation. However, a firm is nothing more than a specific way to organize economic activities. One can, therefore, apply the same logic to the analysis of the organization of such an entity. What is a firm's comparative advantage? How should it organize its internal activities? Where are a firm's boundaries? Should it produce a component itself (insourcing) or buy it in the market (outsourcing)?

Significance for Law
The backbone of the economy is its legal structure. Laws define rights and obligations. Contract law, for example, defines the types of contracts that can be used to organize economic activities, and competition law regulates firm behavior, to name only two fields. The legal structure of an economy, therefore, promotes, or constrains specialization and trade. Hence, one can ask how laws influence economic activities.

2.2.2.2 Solutions to Exercise 2

1. The production-possibility frontier for $\alpha = 8$:
   Country A: $x_1^A = 1{,}000 - x_2^A$
   Country B: $x_1^B = 100\,\alpha - (\frac{\alpha}{9})\,x_2^B \Rightarrow x_1^B = 800 - \frac{8}{9}\,x_2^B$.
   Both production-possibility frontiers are illustrated in Fig. 2.1.
2. The opportunity costs of good i for a given country, k, in units of good j, are given by $OC_{ij}^k = \left|\frac{dx_j^k}{dx_i^k}\right|$, where $i, j \in \{1, 2\}, i \neq j$, and $k \in \{A, B\}$. The results are given in Table 2.1.
3. Looking for comparative advantages, one has to compare the countries' opportunity costs, which leads to the following question: For what values of α does

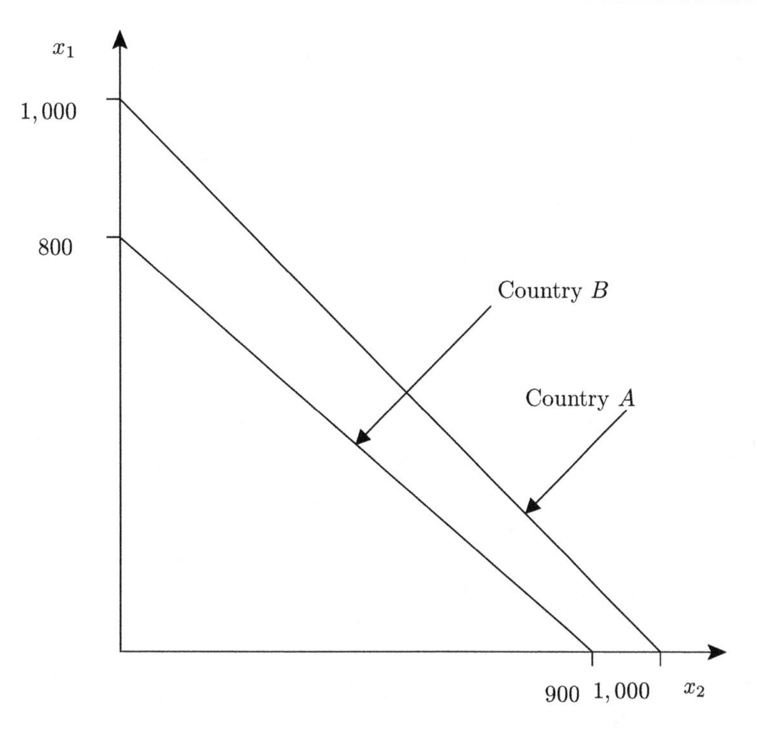

Figure 2.1 Exercise 2.1. The production-possibility frontiers of countries A and B for $\alpha = 8$

one have $OC_{12}^A \left\{ \begin{matrix} > \\ = \\ < \end{matrix} \right\} OC_{12}^B$? By plugging in the results from Question 2, the conditions simplify to

$$OC_{12}^A \left\{ \begin{matrix} > \\ = \\ < \end{matrix} \right\} OC_{12}^B \quad \Leftrightarrow \quad 1 \left\{ \begin{matrix} > \\ = \\ < \end{matrix} \right\} \frac{9}{\alpha} \quad \Leftrightarrow \quad \alpha \left\{ \begin{matrix} > \\ = \\ < \end{matrix} \right\} 9.$$

- If $\alpha > 9$, then country B has a comparative advantage in the production of good 1 and country A in the production of good 2.
- If $\alpha < 9$, then country A has a comparative advantage in the production of good 1 and country B in the production of good 2.
- If $\alpha = 9$, then opportunity costs are identical in both countries and, hence, neither country has a comparative advantage.

4. Neither country has a comparative advantage and there are no gains from trade to be exploited. Whether both countries remain in autarky or trade does not make a difference for the consumption possibilities. If the countries remain in autarky, each country produces exactly as much as it consumes: $x_1^A = x_2^A = 500$ and $x_1^B = x_2^B = 450$.

Table 2.2 Exercise 2.5. Production plans with complete specialization

	Country A	Country B	Total: $x_i^{AB} = x_i^A + x_i^B$
Good 1	$x_1^A = 0$	$x_1^B = 1{,}000$	$x_1^{AB} = 1{,}000$
Good 2	$x_2^A = 1{,}000$	$x_2^B = 0$	$x_2^{AB} = 1{,}000$

Table 2.3 Exercise 2.5. Production plans under autarky

	Country A	Country B	Total: $x_i^{AB} = x_i^A + x_i^B$
Good 1	$x_1^A = 500$	$x_1^B = 500$	$x_1^{AB} = 1{,}000$
Good 2	$x_2^A = 500$	$x_2^B = 450$	$x_2^{AB} = 950$

5. Since $\alpha = 10$, country B has a comparative advantage in the production of good 1. Complete specialization, in the direction of comparative advantage, implies that $x_1^A = 0$, $x_2^A = 1{,}000$, $x_1^B = 1{,}000$, and $x_2^B = 0$. To be more specific, see Table 2.2.
   Assuming that each country produces 500 units of the first good and uses the remaining resources to produce the second good, one gets the following production plans under autarky (see Table 2.3).
   The total production of good 1 remains unchanged by specialization, but the total production of good 2 increases by 50 units. Through trade, these gains in total production can be split between the countries, making both of them potentially better off. What the new allocation with specialization and trade will look like, however, cannot be determined and depends on each country's negotiation power. In addition, one has to make sure that the gains from trade within each country are distributed in a way that each citizen profits or is, at least, not worse off (methodological individualism forces one to think about economic phenomena from the point of view of individual human beings). See Chapter 1.1.

2.2.2.3 Solutions to Exercise 3

1. Assume that the goods are distributed equally among the citizens within each country to abstract from problems of intra-country distribution. The inhabitants of each country, k ($k \in \{A, B, C\}$), maximize their own consumption, subject to the condition that both goods are consumed in equal quantities, $y_1^k = y_2^k$. Under the assumption that, in autarky, production equals consumption (i.e., $y_1^k = x_1^k$ and $y_2^k = x_2^k$), the countries' optimal production plans have the following property: $x_1^k = x_2^k$. The production-possibility frontier (PPF) determines the maximum quantity that can be produced of each good, given the quantity produced of the other good. Applying the constraint $x_1^k = x_2^k$ to the

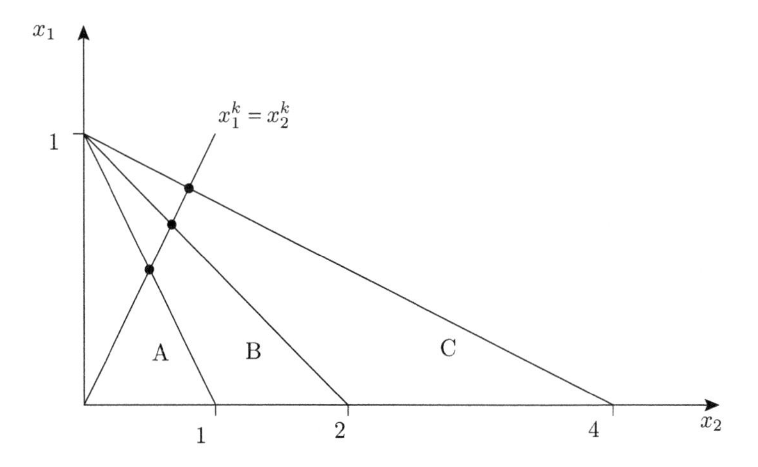

Figure 2.2 Exercise 3.1. The countries' PPFs in autarky and their optimal production plans

PPFs yields the optimal quantities:

$$x_1^A = 1 - x_2^A \wedge x_1^A = x_2^A \quad \Rightarrow x_1^A = \frac{1}{2}, x_2^A = \frac{1}{2}$$
$$x_1^B = 1 - \frac{1}{2}x_2^B \wedge x_1^B = x_2^B \Rightarrow x_1^B = \frac{2}{3}, x_2^B = \frac{2}{3}$$
$$x_1^C = 1 - \frac{1}{4}x_2^C \wedge x_1^C = x_2^C \Rightarrow x_1^C = \frac{4}{5}, x_2^C = \frac{4}{5}.$$

This result is illustrated in Fig. 2.2.

2. If countries A and C establish a free-trade agreement, the countries' joint PPF, as illustrated in Fig. 2.3, includes the points at which each country produces only good 1 or 2, respectively. Starting from point $(x_2, x_1) = (0, 2)$, that is, from the point at which only good 1 is produced, the PPF has a slope of $-\frac{1}{4}$ (the slope of country C's autarky PPF) until it reaches the point $(4, 1)$.
At this point, the slope changes to -1 (the slope of country A's autarky PPF) until the PPF reaches the point at which both countries produce only good 2 (Point $(5, 0)$). The joint PPF is, hence, given as:

$$x_1^{AC} = \begin{cases} 2 - \frac{1}{4}x_2^{AC} & \text{for} \quad 0 \leq x_2^{AC} < 4, \\ 5 - x_2^{AC} & \text{for} \quad 4 \leq x_2^{AC} < 5, \\ 0 & \quad\quad \text{else,} \end{cases} \tag{2.1}$$

where x_i^{AC} is the quantity of good i jointly produced by countries A and C.

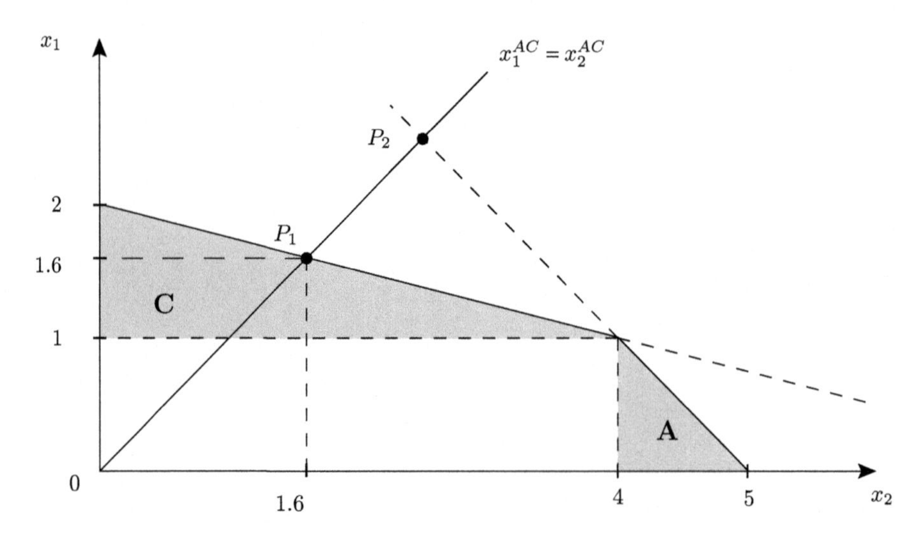

Figure 2.3 Exercise 3.2. Joint PPF of countries *A* and *C*

Optimal production maximizes consumption, subject to the condition that both goods are consumed in equal quantities, $y_1^{AC} = y_2^{AC}$ (which, again, can be transformed into $x_1^{AC} = x_2^{AC}$, since we can still assume that all of the goods that are produced are also consumed). Applying the constraint $x_1^{AC} = x_2^{AC}$ to the joint PPF yields the optimal quantities (see point P_1 in Fig. 2.3):

$$x_1^{AC} = 2 - \frac{1}{4} x_2^{AC} \wedge x_1^{AC} = x_2^{AC} \Rightarrow x_1^{AC} = 1.6, x_2^{AC} = 1.6.$$

Country *A* completely specializes in the production of good 1 ($x_1^A = 1$), whereas country *C* produces 0.6 units of good 1 and 1.6 units of good 2:

$$x_1^A = 1, \qquad x_2^A = 0,$$
$$x_1^C = 0.6, \qquad x_2^C = 1.6.$$

Any piecewise-defined function (like the joint PPF of countries A and C) consists of multiple subfunctions, each of which is paired with an interval (see Eq. 2.1). A common mistake when calculating the optimal production plan of the joint PPF is to not consider the interval to which a specific subfunction applies. In Question 2, this would lead to the following (wrong!) solution (see point P_2 of Fig. 2.3):

$$x_1^{AC} = 5 - x_2^{AC} \wedge x_1^{AC} = x_2^{AC} \Rightarrow x_1^{AC} = 2.5,\ x_2^{AC} = 2.5. \quad ↯$$

a) In autarky, countries *A* and *C* produce 1.3 units of each good in total (see Question 1). Due to the specialization under the free-trade agreement, this quantity increases to 1.6. Hence, there are gains from trade of 0.3 units of each good. If these gains from trade are split equally, each country gains 0.15 units of each good in comparison to what they would have made in autarky. For country B, nothing changes in comparison to Question 1, since it is not part of the trade agreement. Consumption then becomes:

- Country A: $y_1^A = y_2^A = \frac{1}{2} + \frac{3}{20} = \frac{13}{20} = 0.65$,
- Country B: $y_1^B = y_2^B = \frac{2}{3} \approx 0.67$,
- Country C: $y_1^C = y_2^C = \frac{4}{5} + \frac{3}{20} = \frac{19}{20} = 0.95$.

b) Now, the surplus is not split equally. Country C reaps all of the gains from trade alone and country A consumes as much as it would under autarky.

- Country A: $y_1^A = y_2^A = \frac{1}{2} + 0 = 0.5$,
- Country B: $y_1^B = y_2^B = \frac{2}{3} \approx 0.67$,
- Country C: $y_1^C = y_2^C = \frac{4}{5} + \frac{3}{10} = \frac{11}{10} = 1.1$.

3. The new joint PPF is

$$x_1^{ABC} = \begin{cases} 3 - \frac{1}{4}x_2^{ABC} & \text{for} \quad 0 \leq x_2^{ABC} < 4, \\ 4 - \frac{1}{2}x_2^{ABC} & \text{for} \quad 4 \leq x_2^{ABC} < 6, \\ 7 - x_2^{ABC} & \text{for} \quad 6 \leq x_2^{ABC} < 7, \\ 0 & \quad\quad \text{else,} \end{cases} \tag{2.2}$$

where x_i^{ABC} is the quantity of good i jointly produced by countries A, B, and C.

The optimal production plan maximizes the total quantity of both goods, under the constraint that both goods are produced (and consumed) in equal amounts. Applying the constraint $x_1^{ABC} = x_2^{ABC}$ to the joint PPF (see Eq. 2.2) yields the optimal quantities (see point P_1 in Fig. 2.4):

$$x_1^{ABC} = 3 - \frac{1}{4}x_2^{ABC} \wedge x_1^{ABC} = x_2^{ABC} \Rightarrow x_1^{ABC} = 2.4, \, x_2^{ABC} = 2.4.$$

Hence, the new optimal total production plan is $x_1^{ABC} = x_2^{ABC} = 2.4$. Countries A and B completely specialize in the production of good 1 and produce 1 unit each, country C produces 0.4 units of good 1 and 2.4 units of good 2.

$$\begin{aligned} x_1^A &= 1, & x_2^A &= 0, \\ x_1^B &= 1, & x_2^B &= 0, \\ x_1^C &= 0.4, & x_2^C &= 2.4. \end{aligned}$$

Please note once more, not considering the interval to which each subfunction applies leads to the following (wrong!) solutions:

- *Point P_2 in Fig.* 2.4:

$$x_1^{ABC} = 4 - \frac{1}{2}x_2^{ABC} \wedge x_1^{ABC} = x_2^{ABC} \Rightarrow x_1^{AC} \approx 2.67, \, x_2^{AC} \approx 2.67. \quad ↯$$

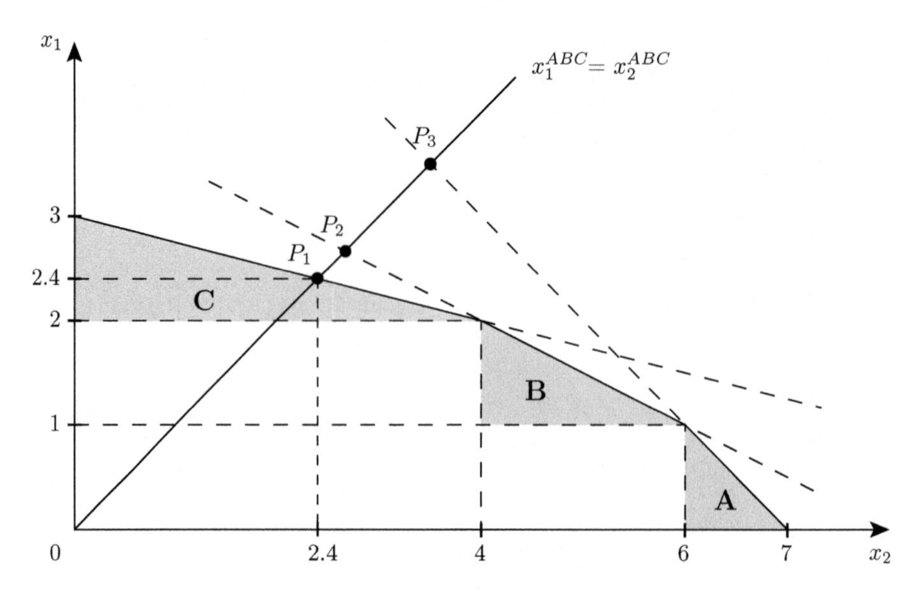

Figure 2.4 Exercise 3.3. Joint PPF of countries A, B, and C

- *Point P_3 in Fig.* 2.4:

$$x_1^{ABC} = 7 - x_2^{ABC} \wedge x_1^{ABC} = x_2^{ABC} \Rightarrow x_1^{ABC} = 3.5,\ x_2^{ABC} = 3.5. \qquad ↯$$

a) In autarky, all countries produce a total of $\frac{59}{30} \approx 1.97$ units of each good (see Question 1), while under the free-trade agreement, production increases to $\frac{72}{30} = 2.4$ units. In comparison to autarky, the surplus amounts to $\frac{72}{30} - \frac{59}{30} = \frac{13}{30} \approx 0.43$ units of each good. If this surplus is split equally between the countries, each country gets $\frac{13}{90} \approx 0.14$:
   - Country A: $y_1^A = y_2^A = \frac{1}{2} + \frac{13}{90} = \frac{58}{90} \approx 0.64$,
   - Country B: $y_1^B = y_2^B = \frac{2}{3} + \frac{13}{90} = \frac{73}{90} \approx 0.81$,
   - Country C: $y_1^C = y_2^C = \frac{4}{5} + \frac{13}{90} = \frac{85}{90} \approx 0.94$.

b) Under the first AC agreement, A and C jointly produce $\frac{48}{30} = 1.6$ units of each good, while B produces $\frac{20}{30} \approx 0.67$. Under the ABC agreement, the total production increases to $\frac{72}{30} = 2.4$. The gains from specialization add up to $\frac{72}{30} - (\frac{48}{30} + \frac{20}{30}) = \frac{4}{30} \approx 0.13$. If countries split the surplus equally, a country's consumption increases by $\frac{4}{90} \approx 0.04$:
   - Country A: $y_1^A = y_2^A = \frac{13}{20} + \frac{4}{90} = \frac{125}{180} \approx 0.69$,
   - Country B: $y_1^B = y_2^B = \frac{2}{3} + \frac{4}{90} = \frac{128}{180} \approx 0.71$,
   - Country C: $y_1^C = y_2^C = \frac{19}{20} + \frac{4}{90} = \frac{179}{180} \approx 0.99$.

   As a result of B joining the trade agreement, compared to Question 2a) total production in the three countries increases by $\frac{4}{30} \approx 0.13$ units of each good in Question 3b). The reason is that country C further specializes in the

Table 2.4 Exercise 3.4. Results

Question	Country A	Country B	Country C
1	$y_i^A = \frac{1}{2}$	$y_i^B = \frac{2}{3}$	$y_i^C = \frac{4}{5}$
2a)	$y_i^A = \frac{13}{20}$	$y_i^B = \frac{2}{3}$	$y_i^C = \frac{19}{20}$
2b)	$y_i^A = \frac{1}{2}$	$y_i^B = \frac{2}{3}$	$y_i^C = \frac{11}{10}$
3a)	$y_i^A = \frac{58}{90}$	$\boldsymbol{y_i^B = \frac{73}{90}}$	$y_i^C = \frac{85}{90}$
3b)	$\boldsymbol{y_i^A = \frac{125}{180}}$	$y_i^B = \frac{128}{180}$	$y_i^C = \frac{179}{180}$
3c)	$y_i^A = \frac{49}{90}$	$y_i^B = \frac{64}{90}$	$\boldsymbol{y_i^C = \frac{103}{90}}$

production of good 2 and country B now fully specializes in the production of good 1.

c) Since total production, under the second AC agreement, equals the total production under the first AC agreement, the gains from specialization, again, add up to $\frac{4}{30}$. However, now the reference point of the equal distribution of the gains from trade differs (in comparison to Question 3b)):
   - Country A: $y_1^A = y_2^A = \frac{1}{2} + \frac{4}{90} = \frac{49}{90} \approx 0.54$,
   - Country B: $y_1^B = y_2^B = \frac{2}{3} + \frac{4}{90} = \frac{64}{90} \approx 0.71$,
   - Country C: $y_1^C = y_2^C = \frac{11}{10} + \frac{4}{90} = \frac{103}{90} \approx 1.14$.

4. Table 2.4 summarizes the results of Exercise 3, where each country's maximal consumption y_i of good i ($i = 1, 2$) is highlighted.
In the first ABC trade agreement (Question 3a)), the benchmark for calculating and distributing the gains from trade is the consumption under autarky. Clearly, everybody gains from this form of market integration (simultaneous integration), i.e. if we assume that no AC trade agreement had existed before. However, in the context of sequential integration, i.e. if we assume that the AC agreement had existed before, this no longer is true.
Countries A and C are worse off under the first ABC agreement (see Question 3a)) compared to the first AC agreement (see Question 2a)). Under the second ABC trade agreement (see Question 3b)), all countries gain, but now the total gain of the inhabitants in country B decreases in comparison to the first ABC trade agreement. This means that, in order to expand the free-trade agreement in a manner that benefits the inhabitants in all countries, the reference point for calculating and distributing the gains from trade has to be the old AC agreement.
Country C is worse off under the first and second ABC trade agreements (see Questions 3a) and 3b)) compared to the second AC agreement (see Question 2b)), while countries A and B are clearly better off. Only under the third ABC trade agreement (see Question 3c)) are all countries better off. However, in that case, the gains from trade are reduced for the inhabitants of countries A and B.
Based on these results we conclude that both i) the order of market integration as well as ii) the reference point chosen for comparison are crucial for the trading partners' assessment of the advantageousness of a trade agreement.

2.3 Multiple Choice

2.3.1 Problems

2.3.1.1 Exercise 1

The production-possibility frontier of two countries A and B with respect to two goods 1 and 2 are given by

$$x_1^A = 20 - x_2^A \text{ and } x_1^B = (1+\alpha)\left(10 - x_2^B\right),$$

with $\alpha \geq 0$ (x is used to denote the quantity produced). There lives exactly one individual in each country who has to consume exactly 10 units of good 1, $y_1^A = y_1^B = 10$ (y is used to denote consumption). The rest of consumption accounts for good 2 (y_2^A, y_2^B).

1. What is the consumption of good 2 in country A (y_2^A) and B (y_2^B) in autarky?
   a) $y_2^A = 0$ and $y_2^B = \alpha$.
   b) $y_2^A = 20$ and $y_2^B = \frac{10\alpha}{1+\alpha}$.
   c) $y_2^A = 0$ and $y_2^B = \frac{\alpha}{1+\alpha}$.
   d) $y_2^A = 10$ and $y_2^B = \frac{10\alpha}{1+\alpha}$.
   e) None of the above answers are correct.
2. What is the critical value of α from which country B has a comparative advantage in the production of good 2?
   a) $\alpha = 0$.
   b) $\alpha = 1$.
   c) Never, as it would imply that $\alpha < 0$, which is excluded by assumption.
   d) $\alpha > 1$.
   e) None of the above answers are correct.
3. What is the critical value of α from which country B has an absolute advantage in the production of good 1?
   a) $\alpha > 1$.
   b) $\alpha = 0$.
   c) Never, as it would imply that $\alpha < 0$, which is excluded by assumption.
   d) $\alpha = 1$.
   e) None of the above answers are correct.

Assume that countries A and B concluded a trade agreement and that both countries determine their production so as to maximize the consumption of good 2 under the constraint that $y_1^A = y_1^B = 10$. Assume further that $\alpha = 1$.

4. Determine the production plans that maximize the gains from trade.
   a) $x_1^A = 20$, $x_2^A = 0$, $x_1^B = 0$, $x_2^B = 20$.
   b) $x_1^A = 0$, $x_2^A = 20$, $x_1^B = 20$, $x_2^B = 10$.
   c) $x_1^A = 0$, $x_2^A = 20$, $x_1^B = 20$, $x_2^B = 0$.
   d) $x_1^A = 10$, $x_2^A = 10$, $x_1^B = 20$, $x_2^B = 0$.
   e) None of the above answers are correct.

5. Determine the gains from trade compared to autarky (in units of good 2).
   a) $y_2 = 6$.
   b) $y_2 = 8$.
   c) $y_2 = 5$.
   d) $y_2 = 2$.
   e) None of the above answers are correct.

2.3.1.2 Exercise 2

Assume that there are two goods, 1 and 2, that can be produced in two countries, A and B. In country A, at maximum, a worker can produce 50 units of the first good or 50 units of the second good or any linear combination of these two quantities, per annum. In country B, at maximum, a worker can produce $\theta > 0$ units of the first good or 75 units of the second good, or any linear combination of these two quantities per year. Each country has 200 workers. Every individual wants to always consume both goods in equal quantities.

1. Determine the opportunity costs of good 1 in units of good 2 for both countries.
   a) The opportunity costs of good 1 in country A are 2, in country B they are $\frac{75}{\theta}$.
   b) The opportunity costs of good 1 in country A are 2, in country B they are $75\,\theta$.
   c) The opportunity costs of good 1 in country A are 1, in country B they are $\frac{75}{\theta}$.
   d) The opportunity costs of good 1 in country A are 1, in country B they are $75\,\theta$.
   e) None of the above answers are correct.

Now, assume that $\theta = 25$.

2. Determine the production possibilities for both countries, as well as the comparative advantages.
   a) The production-possibility frontier of country A is $x_1^A = 10{,}000 - x_2^A$, the production-possibility frontier of country B is $x_1^B = 5{,}000 - x_2^B$. Country A has a comparative advantage in the production of good 1, country B in the production of good 2.
   b) The production-possibility frontier of country A is $x_1^A = 10{,}000 - x_2^A$, the production-possibility frontier of country B is $x_1^B = 5{,}000 - \frac{1}{3}x_2^B$. Country A has a comparative advantage in the production of good 2, country B in the production of good 1.
   c) The production-possibility frontier of country A is $x_1^A = 10{,}000 - x_2^A$, the production-possibility frontier of country B is $x_1^B = 5{,}000 - \frac{1}{3}x_2^B$. Country A has a comparative advantage in the production of good 1, country B in the production of good 2.
   d) The production-possibility frontier of country A is $x_1^A = 10{,}000 - x_2^A$, the production-possibility frontier of country B is $x_1^B = 5{,}000 - x_2^B$. Neither country has a comparative advantage.
   e) None of the above answers are correct.

3. Determine the autarky consumption plans of the two countries.
   a) Country A consumes 10,000 units of each good, country B consumes 7,500 units of each good.
   b) Country A consumes 7,500 units of each good, country B consumes 4,750 units of each good.
   c) Country A consumes 2,500 units of each good, country B consumes 3,750 units of each good.
   d) Country A consumes 5,000 units of each good, country B consumes 7,500 units of each good.
   e) None of the above answers are correct.
4. Determine the consumption maximizing production plans of the countries if they specialize and trade.
   a) They produce a total of 7,500 units of each good.
   b) They produce a total of 8,125 units of each good.
   c) They produce a total of 12,500 units of each good.
   d) They produce a total of 11,250 units of each good.
   e) None of the above answers are correct.
5. Now, assume that they need an intermediary to conduct trade, who demands a one-time payment of C units of each good in compensation for her services. Determine at which values of C the trade is worthwhile if the net trade profits are split equally between the countries.
   a) Only for $C < 5{,}000$ trade is worthwhile.
   b) Only for $C < 2{,}500$ trade is worthwhile.
   c) Only for $C < 1{,}250$ trade is worthwhile.
   d) Only for $C < 3{,}750$ trade is worthwhile.
   e) None of the above answers are correct.

2.3.1.3 Exercise 3

Assume that the production possibilities of three countries (A, B, and C) regarding goods 1 and 2 are given by

$$x_1^A = 2 - x_2^A, \qquad x_1^B = 2 - \frac{1}{2}x_2^B \qquad \text{and} \qquad x_1^C = (1+\alpha)(1 - x_2^C),$$

where $\alpha \geq 0$. In each country lives one individual that must always consume exactly one unit of good 1, i.e. $y_1^A = y_1^B = y_1^C = 1$. The rest of consumption is allocated to good 2 (y_2^A, y_2^B, or y_2^C). Suppose that there are two possible trading agreements: trading agreement AB (consisting of A and B) and trading agreement ABC (consisting of A, B, and C).

1. How many units of good 2 would the individual consume in country A (y_2^A), the individual in country B (y_2^B), and the individual in country C (y_2^C) in autarky?
   a) $y_2^A = 1,\ y_2^B = 2,\ y_2^C = \frac{\alpha}{1+\alpha}$.
   b) $y_2^A = 1,\ y_2^B = 0,\ y_2^C = \alpha$.
   c) $y_2^A = 2,\ y_2^B = 4,\ y_2^C = \frac{\alpha}{1+\alpha}$.

d) $y_2^A = 1$, $y_2^B = \frac{1}{2}$, $y_2^C = \alpha$.
e) None of the above answers are correct.

2. Assume that country A and B have concluded a trade agreement (AB). In AB each of the two countries A and B specializes in the direction of its comparative advantage in order to maximize consumption in AB. Subsequently the goods are traded. How many units of good 1 are produced by country A (x_1^A) and how many units of good 1 are produced by country B (x_1^B)?
   a) $x_1^A = 2$, $x_1^B = 2$.
   b) $x_1^A = 1$, $x_1^B = \frac{4}{3}$.
   c) $x_1^A = 0$, $x_1^B = 2$.
   d) $x_1^A = 2$, $x_1^B = 0$.
   e) None of the above answers are correct.
3. The trade surplus in the trade agreement AB (compared to autarky) is measured in units of the second good. This trade surplus shall be divided between the two countries equally. How large are y_2^A and y_2^B?
   a) $y_2^A = 1.5$, $y_2^B = 2.5$.
   b) $y_2^A = 2.5$, $y_2^B = 4.5$.
   c) $y_2^A = 1$, $y_2^B = 1$.
   d) $y_2^A = 1.5$, $y_2^B = \frac{11}{6}$.
   e) None of the above answers are correct.
4. Assume that countries A, B, and C have concluded a trade agreement (ABC) and that, in order to maximize consumption in ABC, each of the three countries A, B, and C specializes in producing the one good in which it has a comparative advantage. The trade surplus in trade agreement ABC (compared to autarky) is measured in units of the second good. How large must α be for the trade surplus of ABC to be equal in size to the trade surplus of the trade agreement AB from Question 3?
   a) $\alpha > 0$.
   b) $\alpha = 0$.
   c) That is never the case as trade surplus always increases when an additional trading partner is added.
   d) α cannot be determined.
   e) None of the above answers are correct.

2.3.1.4 Exercise 4

Assume that there are two goods, 1 and 2, that can be produced by two individuals, A and B. With one unit of labor, individual A can produce, at most, 10 units of the first good or, at most, 10 units of the second good or any linear combination of these quantities per year. With one unit of labor, individual B can produce, at most, α units of the first good or, at most, 12 units of the second good or any linear combination of these quantities per year. Each individual has 100 units of time. A and B want to consume the goods 1 and 2 at a ratio of 2 : 1.

1. Determine both individuals' opportunity costs for good 1 in units of good 2.
   a) Individual A's opportunity costs for good 1 are 10 and individual B's opportunity costs for good 1 are α.
   b) Individual A's opportunity costs for good 1 are 10 and individual B's opportunity costs for good 1 are 12.
   c) Individual A's opportunity costs for good 1 are 1 and individual B's opportunity costs for good 1 are $\frac{\alpha}{12}$.
   d) Individual A's opportunity costs for good 1 are 1 and individual B's opportunity costs for good 1 are $\frac{12}{\alpha}$.
   e) None of the above answers are correct.
2. Determine the individuals' comparative advantages.
   a) If $\alpha > 12$, individual A has a comparative advantage in the production of good 1 and individual B in the production of good 2. If $\alpha < 12$, individual A has a comparative advantage in the production of good 2 and individual B in the production of good 1. If $\alpha = 12$, neither individual has a comparative advantage.
   b) If $\alpha < 12$, individual A has a comparative advantage in the production of good 1 and individual B in the production of good 2. If $\alpha > 12$, individual A has a comparative advantage in the production of good 2 and individual B in the production of good 1. If $\alpha = 12$, neither individual has a comparative advantage.
   c) Regardless of α's value, A has a comparative advantage in the production of good 1. Additionally, if $\alpha > 10$, individual A also has a comparative advantage in the production of good 2. Individual B has a comparative advantage in the production of good 2 if $\alpha < 10$. If $\alpha = 10$, neither individual has a comparative advantage in the production of good 2.
   d) If $\alpha > 1$, individual A has a comparative advantage in the production of good 1 and individual B in the production of good 2. If $\alpha < 1$, individual A has a comparative advantage in the production of good 2 and individual B in the production of good 1. If $\alpha = 1$, neither individual has a comparative advantage.
   e) None of the above answers are correct.
3. Determine both individuals' production-possibility frontiers.
   a) Individual A's production-possibility frontier is $x_1^A = 10 - x_2^A$ and individual B's is $x_1^B = \alpha - \frac{\alpha}{12} \cdot x_2^B$.
   b) Individual A's production-possibility frontier is $x_1^A = 10 - x_2^A$ and individual B's is $x_1^B = \alpha - \frac{12}{\alpha} \cdot x_2^B$.
   c) Individual A's production-possibility frontier is $x_1^A = 1{,}000 - x_2^A$ and individual B's is $x_1^B = 100 \cdot \alpha - \frac{12}{\alpha} \cdot x_2^B$.
   d) Individual A's production-possibility frontier is $x_1^A = 1{,}000 - x_2^A$ and individual B's is $x_1^B = 100 \cdot \alpha - \frac{\alpha}{12} \cdot x_2^B$.
   e) None of the above answers are correct.

Now, assume that $\alpha = 6$.

4. Determine the consumption plan for both individuals in autarky.
   a) In autarky, individual A consumes 800 units of good 1 and 400 units of good 2. In autarky, individual B consumes 600 units of good 1 and 300 units of good 2.
   b) In autarky, individual A consumes $666\frac{2}{3}$ units of good 1 and $333\frac{1}{3}$ units of good 2. In autarky, individual B consumes 800 units of good 1 and 400 units of good 2.
   c) In autarky, individual A consumes 800 units of good 1 and 400 units of good 2. In autarky, individual B consumes 480 units of good 1 and 240 units of good 2.
   d) In autarky, individual A consumes $666\frac{2}{3}$ units of good 1 and $333\frac{1}{3}$ units of good 2. In autarky, individual B consumes 480 units of good 1 and 240 units of good 2.
   e) None of the above answers are correct.
5. Determine the individuals' optimal production plans if they can specialize and trade.
   a) With specialization and trade, 1,280 units of good 1 and 640 units of good 2 will be produced.
   b) With specialization and trade, 1,500 units of good 1 and 750 units of good 2 will be produced.
   c) With specialization and trade, 1,350 units of good 1 and 675 units of good 2 will be produced.
   d) With specialization and trade, 1,400 units of good 1 and 700 units of good 2 will be produced.
   e) None of the above answers are correct.
6. Determine the trade surplus that can maximally be obtained through specialization.
   a) Compared to autarky, 200 additional units of good 1 and 100 additional units of good 2 are produced due to specialization.
   b) Compared to autarky, $133\frac{1}{3}$ additional units of good 1 and $66\frac{2}{3}$ additional units of good 2 are produced due to specialization.
   c) Compared to autarky, $166\frac{2}{3}$ additional units of good 1 and $83\frac{1}{3}$ additional units of good 2 are produced due to specialization.
   d) Compared to autarky, 150 additional units of good 1 and 75 additional units of good 2 are produced due to specialization.
   e) None of the above answers are correct.

2.3.2 Solutions

2.3.2.1 Solutions to Exercise 1

- Question 1, answer d) is correct.
- Question 2, answer c) is correct.
- Question 3, answer a) is correct.
- Question 4, answer c) is correct.
- Question 5, answer c) is correct.

2.3.2.2 Solutions to Exercise 2

- Question 1, answer c) is correct.
- Question 2, answer c) is correct.
- Question 3, answer e) is correct. Country A consumes 5,000 units of each good and country B consumes 3,750 units of each good.
- Question 4, answer d) is correct.
- Question 5, answer b) is correct.

2.3.2.3 Solutions to Exercise 3

- Question 1, answer a) is correct.
- Question 2, answer d) is correct.
- Question 3, answer a) is correct.
- Question 4, answer b) is correct.

2.3.2.4 Solutions to Exercise 4

- Question 1, answer d) is correct.
- Question 2, answer b) is correct.
- Question 3, answer d) is correct.
- Question 4, answer d) is correct.
- Question 5, answer a) is correct.
- Question 6, answer b) is correct.

3 Markets and Institutions – Introduction

3.1 True or False

3.1.1 Statements

3.1.1.1 Block 1

1. A market with few consumers and many suppliers is called a restricted monopsony.
2. A market with one consumer and one supplier is called a bilateral monopoly.
3. A market with many suppliers and many consumers is always a polypoly.
4. A market with few suppliers and one consumer is called a restricted monopoly.

3.1.1.2 Block 2

1. There are more suppliers in an oligopoly than in a restricted monopsony.
2. There are more suppliers in a bilateral oligopoly than in a monopsony.
3. The only difference between a oligopsony and an oligopoly is the number of consumers.
4. The number of suppliers in a bilateral oligopoly is always smaller than in an oligopsony.

3.1.1.3 Block 3

1. The market for airplane travel is an example of an oligopoly.
2. There is exactly one consumer in a bilateral monopoly, a restricted monopsony, and a monopsony.
3. The number of the market participants in a monopoly is always at least as large as in a bilateral monopoly.
4. The car industry is an example of a market with monopolistic competition.

M. Kolmar, M. Hoffmann, *Workbook for Principles of Microeconomics*,
Springer Texts in Business and Economics, https://doi.org/10.1007/978-3-319-62662-8_3

3.1.2 Solutions

3.1.2.1 Sample Solutions for Block 1

1. **False**. A market with few consumers and many suppliers is called an oligopsony. See Chapter 3.
2. **True**. This is true by definition. See Chapter 3.
3. **True**. This is true by definition. See Chapter 3.
4. **False**. Restricted monopolies have one supplier and few consumers. See Chapter 3.

3.1.2.2 Sample Solutions for Block 2

1. **False**. Both market forms have few suppliers. See Chapter 3.
2. **False**. There are few suppliers in a bilateral oligopoly, while there are many suppliers in a monopsony. See Chapter 3.
3. **False**. An oligopsony has many suppliers and few consumers, while an oligopoly has few suppliers and many consumers. See Chapter 3.
4. **True**. Bilateral oligopolies have few suppliers, while oligopsonies have many suppliers. See Chapter 3.

3.1.2.3 Sample Solutions for Block 3

1. **True**. In an oligopoly, there are few suppliers and many consumers. See Chapter 3.
2. **True**. This is true by definition. See Chapter 3.
3. **True**. A bilateral monopoly only has one supplier and one consumer. A monopoly, too, only has one supplier, but many consumers. See Chapter 3.
4. **True**. Consider the discussion of the market for sports-utility vehicles (SUVs) in Chapter 3.

4 Supply and Demand

4.1 True or False

4.1.1 Statements

4.1.1.1 Block 1

1. When income increases, the demanded quantity of an ordinary good decreases.
2. If the price of an inferior good increases, then the demand decreases.
3. Two goods are substitutes to each other if the demand for each good decreases when the price of the other good increases.
4. The demand for a good increases as its price increases. Hence, it is a Giffen good.

4.1.1.2 Block 2

Assume that the market for cars is perfectly competitive. Figure 4.1 shows the relevant market for the car firm CarMaker. For the following questions, please take the demand x_2, the supply y_2 and the equilibrium in point a as reference point.

1. A rival firm, which produces a substitute for the cars by CarMaker, lowers the price of its cars. The new equilibrium is at a point such as i.
2. Due to a process of innovation, CarMaker can reduce the marginal costs. The new equilibrium is at a point such as h.
3. The government increases the motorway toll. The new equilibrium is at a point such as d.
4. The government reduces the mineral oil tax. The new equilibrium is at a point such as f.

4.1.1.3 Block 3

Assume that the market for corn is perfectly competitive. Supply is increasing and demand is decreasing in price.

M. Kolmar, M. Hoffmann, *Workbook for Principles of Microeconomics*, Springer Texts in Business and Economics, https://doi.org/10.1007/978-3-319-62662-8_4

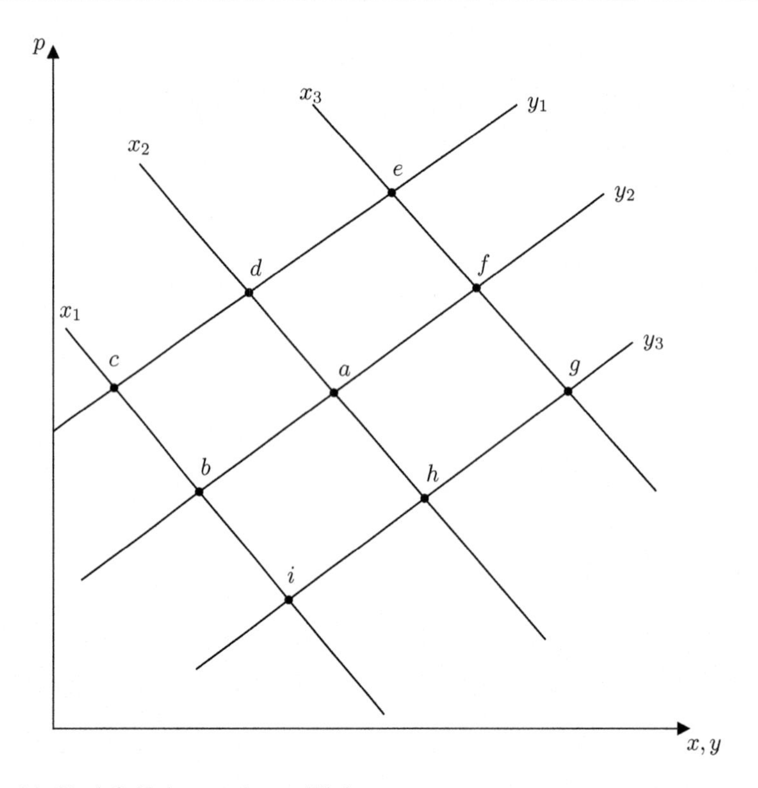

Figure 4.1 Block 2. Various market equilibria

1. A large amount of the crop is destroyed by storms. The equilibrium price thus increases, *ceteris paribus*.
2. All harvesters' wages decrease. The market supply function shifts, *ceteris paribus*, to the left.
3. A new technology allows the production of gasoline from corn. The equilibrium demand for corn decreases and the equilibrium quantity increases, *ceteris paribus*.
4. The income of the consumers of corn increases. The equilibrium price for corn thus increases, *ceteris paribus*.

4.1.2 Solutions

4.1.2.1 Sample Solutions for Block 1

1. **False**. If the price of an ordinary good increases, then its demand decreases. See Definition 4.1 in Chapter 4.2.
2. **False**. The demand for an inferior good decreases if the budget increases. See Definition 4.4 in Chapter 4.2.

3. **False**. Two goods are substitutes if the demand for one good increases when the price for the other good increases. See Definition 4.5 in Chapter 4.2.
4. **True**. See Definition 4.2 in Chapter 4.2.

4.1.2.2 Sample Solutions for Block 2

1. **False**. If the price of the substitute product decreases, the demand for CarMaker's product decreases (see Definition 4.5 in Chapter 4.2). The supply does not change. The new equilibrium is at a point such as b.
2. **True**. By decreasing the marginal costs, the supply function shifts to the right (see Chapter 4.2). Now, the supply increases for each price.
3. **False**. The motorway toll is a complementary good to CarMaker's product. If the complementary good's price increases, the demand for CarMaker's product decreases (see Definition 4.6 in Chapter 4.2). The new equilibrium is in a point like b.
4. **True**. Mineral oil is a complementary good to CarMaker's product. If the complementary good's price decreases, the demand for CarMaker's product increases (see Definition 4.6 in Chapter 4.2).

4.1.2.3 Sample Solutions of Block 3

1. **True**. The supply function shifts to the left, while the demand function remains constant. See Chapter 4.4.
2. **False**. The marginal production costs decrease and, thus, the supply function shifts to the right. See Chapter 4.4.
3. **False**. The demand for corn increases (the demand function shifts to the right) and the equilibrium quantity increases. See Chapter 4.4.
4. **False**. There is not enough information to make this statement. Therefore, it is incorrect. A counterexample: See Chapter 4.4 and Definition 4.4 in Chapter 4.2: A good, i, is called inferior, at given prices and budgets, if the demand for that good, x_i^j, decreases with an increasing budget, b^j.

4.2 Open Questions

4.2.1 Problems

4.2.1.1 Exercise 1

In an (crude) oil producing region, a political conflict emerges that decreases the global aggregate oil supply.

1. What will be the likely consequences for the demand for cars? Illustrate your answer in a price-quantity diagram.
2. How will the decrease in the global oil supply most likely impact the demand for alternative energies?

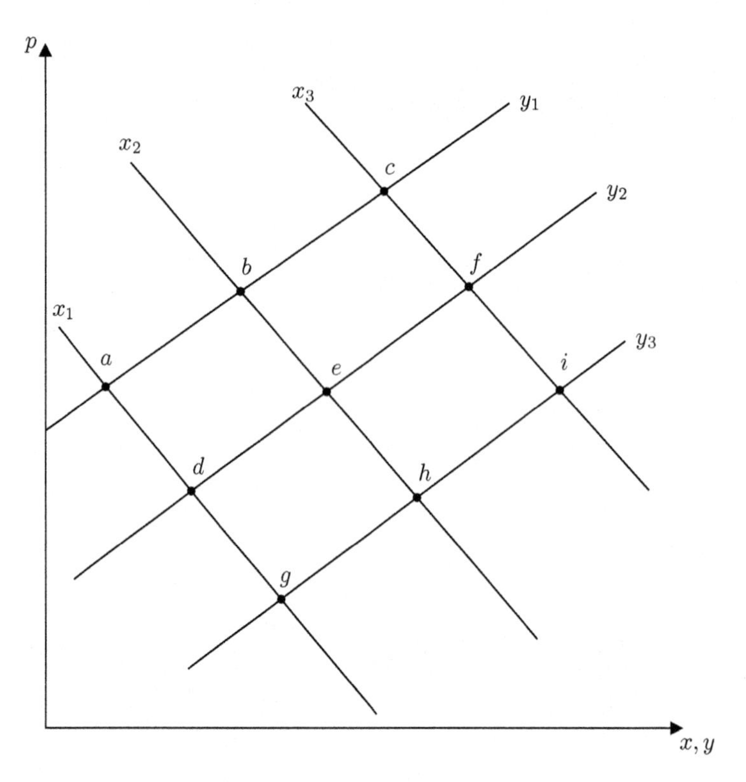

Figure 4.2 Exercise 2. Various market equilibria

4.2.1.2 Exercise 2

Figure 4.2 illustrates the (competitive) market for Dutch cheese. The current equilibrium is represented by point e, while y_2 corresponds to the current supply function and x_2 to the current demand function. What consequences will the following incidents have on the market equilibrium?

1. During wage negotiations, the trade union of the cheese dairy employees succeeds in having the salaries raised.
2. Assume that cheese is a normal good. A recession reduces the consumers' disposable income.
3. According to a scientific study conducted by a reputable independent institute, the consumption of cheese has a positive effect on intellectual abilities.
4. A mysterious epidemic reduces the milk-producing livestock used in the dairy production by half.

4.2.1.3 Exercise 3

Urs and Heidi love ice cream. At a price of 0 Swiss Francs per scoop, Urs demands 10 scoops a *week*. When the price per scoop rises by 1 Swiss Franc, the quantity he demands falls by 2 scoops a week. Heidi's demand at a price of 0 Swiss Francs is 2 scoops of ice cream per *day*. As the price rises by 1 Swiss Franc per scoop, the quantity she demands falls by 1 scoop per *day*. Assume that Urs' and Heidi's demands can be represented by linear demand functions.

1. Calculate Urs' and Heidi's individual *weekly* demand and the aggregate market demand per *week* and illustrate the results using a price-quantity table. Determine the mathematical function that represents market demand.
2. Is ice cream an ordinary good for Urs and Heidi?
3. The local ice cream parlor, iScream, is not willing to offer ice cream for a price of 0 Swiss Francs per scoop. As the price rises by 1 Swiss Franc per scoop, iScream increases supply by 15 scoops of ice cream *per week*. Assume that iScream's supply can be represented by a linear supply function.
   a) Determine the ice cream parlor's supply using a price-quantity table and find the mathematical function that represents market supply.
   b) Draw a supply and demand diagram and determine the market equilibrium.
   c) How does the equilibrium change when Urs' brother Alexander starts to always buy the same amount of ice cream as his brother?

4.2.1.4 Exercise 4

Aggregate demand in a competitive market is given by:

$$x(p) = \begin{cases} 400 - 2p & \text{for } 200 > p \geq 50, \\ 500 - 4p & \text{for } 50 > p. \end{cases}$$

Aggregate supply is given by:

$$y(p) = \begin{cases} -30 + 2p & \text{for } p \geq 20, \\ -10 + p & \text{for } 20 > p \geq 10. \end{cases}$$

$x(p)$ consists of two individual (linear) demand functions for individual A and B ($x^A(p)$ and $x^B(p)$), while $y(p)$ consists of two individual (linear) supply functions for firm C and D ($y^C(p)$ and $y^D(p)$).

1. Analyze the aggregate demand and determine the individual demand functions of individual A and B. Draw the aggregate and both individual demand functions in one diagram.
2. Analyze the aggregate supply and determine the individual supply functions of firm C and D. Draw the aggregate and both individual supply functions in one diagram.
3. Calculate the market price in equilibrium (p^*). What quantity is supplied in equilibrium ($y^* = y(p^*)$)? Illustrate your result in a diagram.
4. What quantity is demanded in equilibrium by individual A and what quantity by individual B? What quantity is supplied in equilibrium by individual C and what quantity by individual D? Illustrate your result in a diagram.

5. Assume that a third firm E enters the market, with the following supply function:

$$y^E(p) = 10\,p.$$

Does this new market entry alter the results of Questions 3 and 4?

4.2.1.5 Exercise 5

Sabine has a monthly income of $b > 0$, which she always entirely spends on n different goods. Note that b may vary from month to month.

1. How many of the n different goods Sabine purchases can maximally be inferior?
2. Let $n = 2$ (good 1 and 2) and let Sabine's demand for good 1 be given by

$$x_1(b, p_1, p_2) = \frac{b}{p_1 + a\,p_2},$$

where p_i represents good i's price and a is a parameter that influences demand.
a) For which values of a are goods 1 and 2 substitutes or complements?
b) Let $a > 0$. Is good 1 inferior or normal? Is it an ordinary or a Giffen good?

4.2.2 Solutions

4.2.2.1 Solutions to Exercise 1

1. Cars and gas are complementary goods. When the gas price increases the demand for cars will fall, assuming that cars are ordinary goods. The demand function shifts to the left (see Fig. 4.3).
2. The demand for alternative energies will increase, because the relative price between oil and alternative energies rises due to supply shortage. Individuals and firms will use alternative energies as substitutes for oil (if this is at all possible).

4.2.2.2 Solutions to Exercise 2

1. The new equilibrium shifts to a new point, such as b.
2. The new equilibrium shifts to a new point, such as d.
3. The new equilibrium shifts to a new point, such as f.
4. The new equilibrium shifts to a new point, such as b.

4.2.2.3 Solutions to Exercise 3

1. At a price of 0 Swiss Francs per scoop ($p = 0$), Urs demands 10 scoops per week. His demand decreases by 2 scoops if the price rises by 1 Swiss Franc. Since his demand is a linear function, it is:

$$x^U(p) = 10 - 2\,p \text{ for } p \leq 5.$$

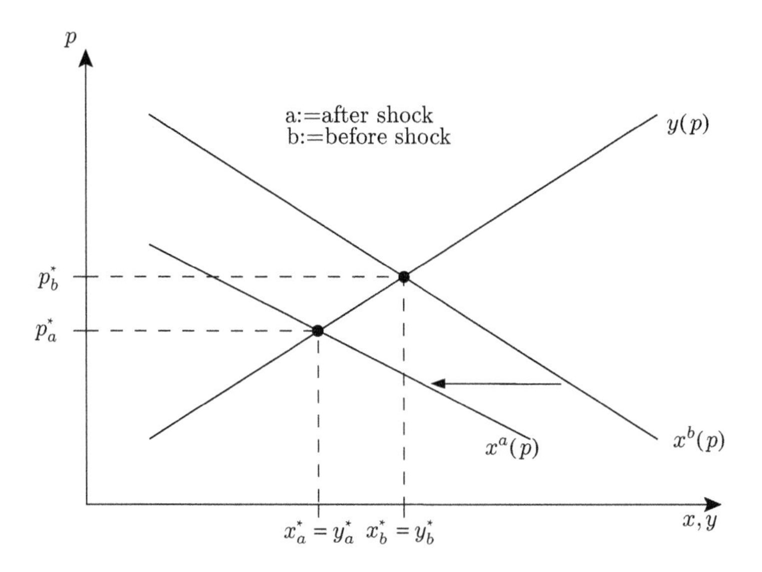

Figure 4.3 Exercise 1.1. Supply and demand before and after the shock

Heidi demands 2 scoops per day if $p = 0$. This implies that, at that price, she demands 14 scoops per week. If the price increases by 1 Swiss Franc, her demand falls by 1 scoop per day, or 7 scoops per week, respectively. If $p = 1$, then Heidi's weekly demand is 7 scoops. As her demand is a linear function, her weekly consumption would be:

$$x^H(p) = 14 - 7\,p \text{ for } p \leq 2.$$

Table 4.1 illustrates the individual and total demand per week.
The function of market demand is:

$$x(p) = \begin{cases} 24 - 9\,p & \text{for} \quad p \leq 2, \\ 10 - 2\,p & \text{for} \quad 2 < p \leq 5, \end{cases}$$

where $x(p) \equiv x^H(p) + x^U(p)$.

2. An ordinary good implies that demand for the good decreases if its price increases (see Definition 4.1 in Chapter 4.2). Urs' demand falls by 2 scoops per week and Heidi's demand falls by 1 scoop per day if the scoop price rises by 1 Swiss Franc. Clearly, ice cream is an ordinary good for both Urs and Heidi.
3. a) Table 4.2 illustrates the parlor's supply per week contingent on p. The function of market supply is:

   $$y(p) = 15\,p.$$

   b) In equilibrium, supply must equal demand, i.e. $y(p^*) = x(p^*)$. In order to find the equilibrium, we need a case-by-case analysis (see Fig. 4.4):

Table 4.1 Exercise 3.1. Individual and market demand per week

Price	Urs' demand	Heidi's demand	Market demand
0	10	14	24
0.5	9	10.5	19.5
1	8	7	15
1.5	7	3.5	10.5
2	6	0	6
2.5	5	0	5
3	4	0	4
3.5	3	0	3
4	2	0	2
4.5	1	0	1
5	0	0	0

Table 4.2 Exercise 3.3a). Market supply per week

Price	Market supply
0	0
0.5	7.5
1	15
1.5	22.5
2	30
2.5	37.5
3	45

- 1st case P_1: $2 < p \leq 5$

$$\begin{aligned} x(p) &= y(p) \\ \Leftrightarrow 10 - 2p &= 15p \\ \Leftrightarrow p &= \frac{10}{17} \approx 0.59. \qquad ↯ \end{aligned}$$

- 2nd case P_2: $p \leq 2$

$$\begin{aligned} x(p) &= y(p) \\ \Leftrightarrow 24 - 9p &= 15p \\ \Leftrightarrow p &= 1. \qquad \checkmark \end{aligned}$$

Equilibrium price and quantity are $p^* = 1$ and $x^* = y^* = 15$, respectively.

c) Table 4.3 illustrates the individual and total demand per week contingent on p. The supply function is unchanged. The demand function is now:

$$x(p) = \begin{cases} 34 - 11p & \text{for} \quad p \leq 2, \\ 20 - 4p & \text{for} \quad 2 < p \leq 5, \end{cases}$$

where $x(p) \equiv x^H(p) + x^U(p) + x^A(p)$ (Fig. 4.5).

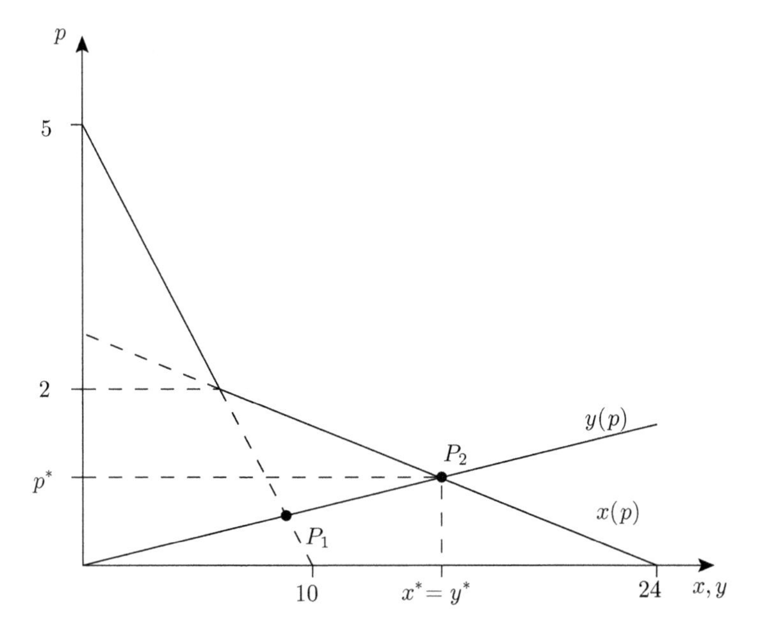

Figure 4.4 Exercise 3.3b). Market supply and demand in equilibrium

In equilibrium, supply must equal demand, i.e. $y(p^*) = x(p^*)$. We distinguish the following cases:

- 1st case P_1: $2 < p \leq 5$

$$\begin{aligned} x(p) &= y(p) \\ \Leftrightarrow 20 - 4p &= 15p \\ \Leftrightarrow p &= \frac{20}{19} \approx 1.05. \quad \lightning \end{aligned}$$

Table 4.3 Exercise 3.3c). Individual demands and market demand

Price	Urs' demand	Heidi's demand	Alexander's demand	Market demand
0	10	14	10	34
0.5	9	10.5	9	28.5
1	8	7	8	23
1.5	7	3.5	7	17.5
2	6	0	6	12
2.5	5	0	5	10
3	4	0	4	8
3.5	3	0	3	6
4	2	0	2	4
4.5	1	0	1	2
5	0	0	0	0

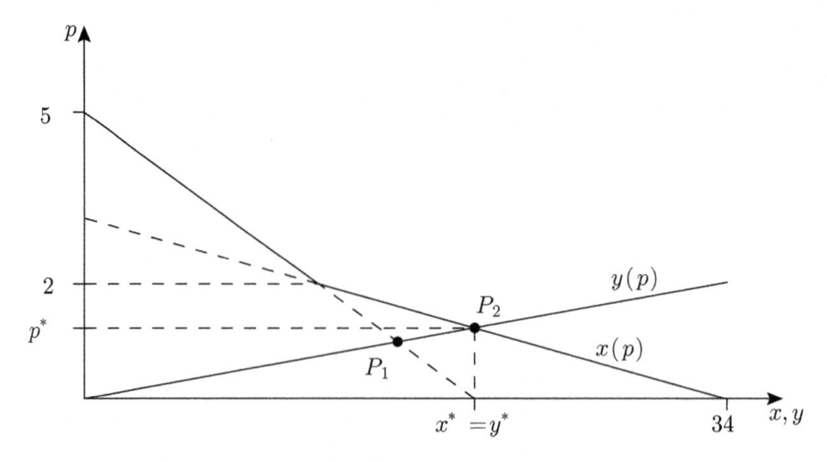

Figure 4.5 Exercise 3.3c). Market supply and demand in equilibrium

- 2nd P_2: $p \leq 2$

$$\begin{aligned} x(p) &= y(p) \\ \Leftrightarrow 34 - 11p &= 15p \\ \Leftrightarrow p &= \frac{17}{13} \approx 1.31. \qquad \checkmark \end{aligned}$$

Supply and demand intersect at a price lower than 2. Thus, equilibrium price and quantity are $p^* = \frac{17}{13} \approx 1.31$ and $x^* = y^* = \frac{255}{13} \approx 19.62$, respectively. The price rises by $\frac{4}{13} \approx 0.31$ Swiss Francs, the quantity increases by $\frac{60}{13} \approx 4.62$ scoops.

4.2.2.4 Solutions to Exercise 4

1. The market demand equals the sum of both individuals' demand functions, i.e.

$$x(p) = x^A(p) + x^B(p). \tag{4.1}$$

The market demand function $x(p)$ has a kink at $p = 50$. Below (above) $p = 50$, the market demand function proceeds flatter (steeper) than above (below) that price (Fig. 4.6).
Apparently, one individual (say A) has a relatively high willingness to pay and is willing to pay as much as 200, while B's maximum willingness to pay is 50. Hence, individual A's demand is:

$$x^A(p) = 400 - 2\,p \text{ for } p \leq 200.$$

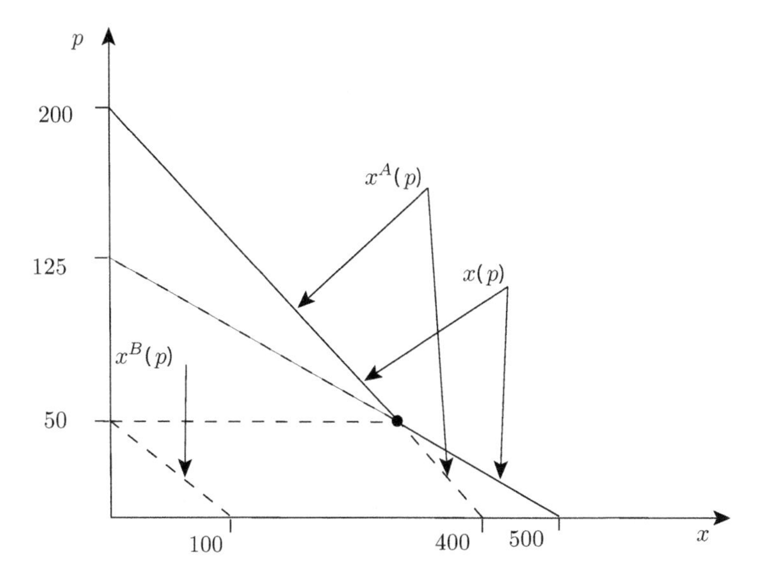

Figure 4.6 Exercise 4.1. Demand

Utilizing Eq. 4.1 gives B's individual demand:

$$\begin{aligned} & x(p) = x^A(p) + x^B(p) \\ \Leftrightarrow \quad & x^B(p) = 500 - 4\,p - (400 - 2\,p) \\ \Leftrightarrow \quad & x^B(p) = 100 - 2\,p \text{ for } p \leq 50. \end{aligned}$$

Total and individual demand are illustrated in Fig. 4.6.

2. The market supply equals the sum of both firms' supply functions, i.e.

$$y(p) = y^C(p) + y^D(p). \tag{4.2}$$

The market supply function $y(p)$ has a kink at $p = 20$. Below (above) $p = 20$, the market supply function proceeds steeper (flatter) than above (below) that price.

Apparently, one firm (say C) is not willing to supply the good for prices below $p = 20$, while firm D will even settle for prices between 10 and 20. Hence, firm D's supply function is:

$$y^D(p) = -10 + p \text{ for } p \geq 10.$$

Utilizing Eq. 4.2 gives C's individual supply function:

$$\begin{aligned} & y(p) = y^C(p) + y^D(p) \\ \Leftrightarrow \quad & y^C(p) = -30 + 2\,p - (-10 + p) \\ \Leftrightarrow \quad & y^C(p) = -20 + p \text{ for } p \geq 20. \end{aligned}$$

Total and individual supply are illustrated in Fig. 4.7.

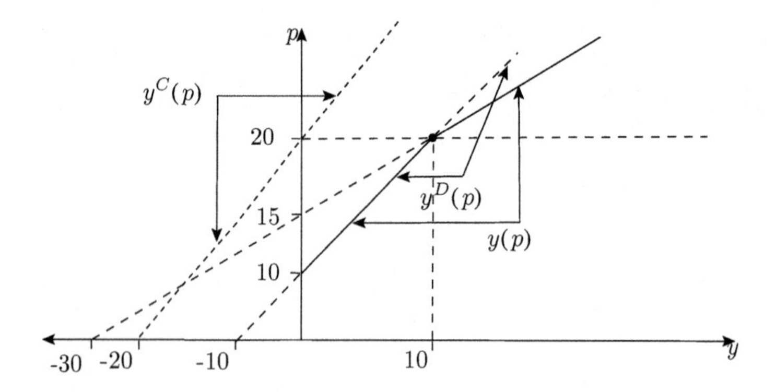

Figure 4.7 Exercise 4.2. Supply

3. To answer this question, we need a case-by-case analysis:
   - 1st case: $10 \leq p < 20$:

$$\begin{aligned} -10 + p &= 500 - 4\,p \\ \Leftrightarrow \qquad 5\,p &= 510 \\ \Leftrightarrow \qquad p &= 102. \qquad ↯ \end{aligned}$$

   - 2nd case: $20 \leq p < 50$:

$$\begin{aligned} -30 + 2\,p &= 500 - 4\,p \\ \Leftrightarrow \qquad 6\,p &= 530 \\ \Leftrightarrow \qquad p &= 88\frac{1}{3}. \qquad ↯ \end{aligned}$$

   - 3rd case: $50 \leq p < 200$:

$$\begin{aligned} -30 + 2\,p &= 400 - 2\,p \\ \Leftrightarrow \qquad 4\,p &= 430 \\ \Leftrightarrow \qquad p &= 107.5. \qquad \checkmark \end{aligned}$$

   The equilibrium market price is $p^* = 107.5$ and the equilibrium market supply is

$$y^* = y(p^*) = -30 + 2\,p^* = 185.$$

   The result is illustrated in Fig. 4.8.

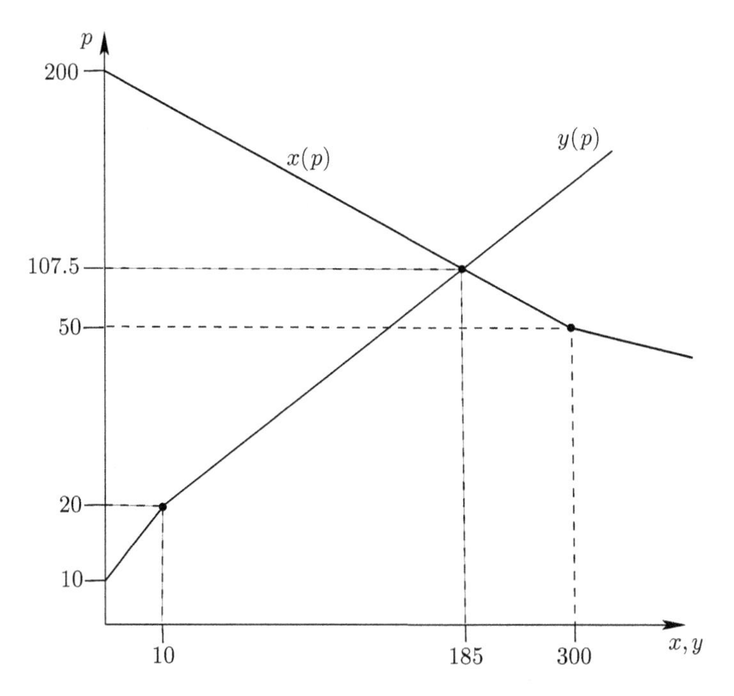

Figure 4.8 Exercise 4.3. Equilibrium

4. Individual demand and individual supply in equilibrium are given by:
   - $x^A(p^*) = 400 - 2\,p^* = 185$,
   - $x^B(p^*) = 0$,
   - $y^C(p^*) = -20 + p^* = 87.5$,
   - $y^D(p^*) = -10 + p^* = 97.5$.
5. The new market supply function ($\tilde{y}(p)$) is given by:

$$\tilde{y}(p) = y(p) + y^E(p) = \begin{cases} -30 + 12\,p, & \text{for } p \geq 20, \\ -10 + 11\,p, & \text{for } 20 > p \geq 10, \\ 10\,p, & \text{for } 10 > p. \end{cases}$$

To find the new equilibrium we, again, need a case-by-case analysis:

- 1st case: $0 \leq p < 10$:

$$\begin{aligned} 10\,p &= 500 - 4\,p \\ \Leftrightarrow \quad 14\,p &= 500 \\ \Leftrightarrow \quad p &= 35\frac{5}{7}. \quad ↯ \end{aligned}$$

- 2nd case: $10 \le p < 20$:

$$\begin{aligned} -10 + 11\,p &= 500 - 4\,p \\ \Leftrightarrow \quad 15\,p &= 510 \\ \Leftrightarrow \quad p &= 34. \quad ↯ \end{aligned}$$

- 3rd case: $20 \le p < 50$:

$$\begin{aligned} -30 + 12\,p &= 500 - 4\,p \\ \Leftrightarrow \quad 16\,p &= 530 \\ \Leftrightarrow \quad p &= 33\frac{1}{8}. \quad \checkmark \end{aligned}$$

The new equilibrium market price is $\tilde{p}^* = 33\frac{1}{8}$, and the equilibrium market supply is[1]

$$\tilde{y}^* = \tilde{y}(\tilde{p}^*) = -30 + 12\,\tilde{p}^* = 367.5.$$

The result is illustrated in Fig. 4.9.
Individual demand and supply in equilibrium are given by:

- $x^A(\tilde{p}^*) = 400 - 2\,\tilde{p}^* = 333.75$,
- $x^B(\tilde{p}^*) = 100 - 2\,\tilde{p}^* = 33.75$,
- $y^C(\tilde{p}^*) = -20 + \tilde{p}^* = 13\frac{1}{8}$,
- $y^D(\tilde{p}^*) = -10 + \tilde{p}^* = 23\frac{1}{8}$,
- $y^E(\tilde{p}^*) = 10\,\tilde{p}^* = 331.25$.

4.2.2.5 Solutions to Exercise 5

1. A good is inferior if its demand falls when income increases. If all goods purchased were inferior, Sabine's expenditures for each good would be falling as her income increases. Hence, total expenditures would also be falling. However, this cannot happen, because Sabine always spends her entire income and, thus, total expenditures must rise as income increases. Hence, there has to be at least one normal good, and there can be at most $n - 1$ inferior goods.
2. a) Good 1 is a substitute for good 2 if the demand for good 1 falls when the price of good 2 falls. Good 1 is a complement to good 2 if the demand for good 1 rises when the price of good 2 falls. The derivative of demand $x_1(p_1, p_2)$ with respect to p_2 is:

$$\frac{\partial x_1(b, p_1, p_2)}{\partial p_2} = \frac{-a\,b}{(p_1 + a\,p_2)^2}$$

[1] Note that we did not analyze the fourth case, where prices run between 50 and 200. The reason for this is simple. We are looking for the intersection of a monotonically decreasing and continuous function ($x(p)$) and a monotonically increasing and continuous function ($\tilde{y}(p)$). Thus, a unique solution must exist, which we already found in the interval $20 \le p < 50$.

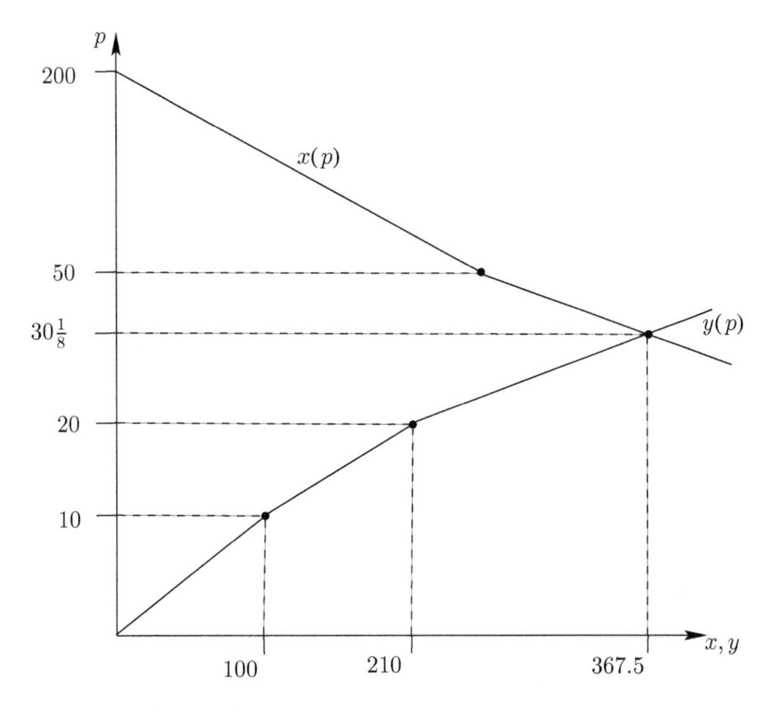

Figure 4.9 Exercise 4.5. Equilibrium

Since we can assume that $b > 0$,

- the derivative is positive for all $a < 0$ and good 1 is a substitute for good 2,
- the derivative is negative for all $a > 0$ and good 1 is a complement to good 2.

b) The derivative of the demand function with respect to b is:

$$\frac{\partial x_1(b, p_1, p_2)}{\partial b} = \frac{1}{p_1 + a\, p_2}.$$

Since we can assume that at least one price is larger than 0, the derivative is positive for all $a > 0$. Hence, demand rises as income increases, and the good is normal.
The derivative of the demand function with respect to p_1 is:

$$\frac{\partial x_1(b, p_1, p_2)}{\partial p_1} = -\frac{b}{(p_1 + a\, p_2)^2}.$$

The derivative is negative for all $a > 0$. Hence, demand decreases as the price increases, and the good is ordinary.

5 Normative Economics

5.1 True or False

5.1.1 Statements

5.1.1.1 Block 1

1. An allocation of given quantities of goods and services is defined as efficient in consumption if it is not possible to reallocate the resources in such a way as to increase the production of one good without reducing the production of another good.
2. An allocation of given quantities of resources is defined as efficient in production if it is not possible to increase the well-being of at least one individual without reducing the well-being of another individual.
3. An allocation is called efficient in production if it is possible to increase the production of at least one good without reducing the production of some other good by reallocating the given quantities of resources.
4. If it is impossible, by reallocating the given quantities of resources, to improve an individual's well-being without reducing another individual's well-being, the allocation is efficient in consumption.

5.1.1.2 Block 2

1. The First Theorem of Welfare Economics states that every equilibrium in a polypoly maximizes consumer surplus.
2. An equilibrium in a perfectly competitive market is Pareto efficient because it maximizes the sum of consumer and producer surplus.
3. The Second Theorem of Welfare Economics states that, under specific conditions, every Pareto-efficient allocation can be achieved trough the market mechanisms.
4. From the First Theorem of Welfare Economics one can deduce that only the equilibrium in a polypoly maximizes the sum of consumer and producer surplus.

M. Kolmar, M. Hoffmann, *Workbook for Principles of Microeconomics*,
Springer Texts in Business and Economics, https://doi.org/10.1007/978-3-319-62662-8_5

5.1.1.3 Block 3

1. An allocation is Pareto-efficient if it is either efficient in consumption or efficient in production.
2. If an allocation is efficient in production, it is also efficient in consumption.
3. If an allocation is efficient in production, then it is also Pareto-efficient.
4. A Pareto-efficient allocation is always efficient in consumption.

5.1.2 Solutions

5.1.2.1 Sample Solutions for Block 1

1. **False**. An allocation of given quantities of goods and services is efficient in consumption if it is not possible to reallocate the goods and services among the consumers in such a way as to increase the well-being of at least one consumer without reducing the well-being of another consumer. See Definition 5.2 in Chapter 5.1.
2. **False**. An allocation of given quantities of resources is efficient in production if it is not possible to reallocate the resources among the producers in such a way as to increase the production of at least one good without reducing the production of some other good. See Definition 5.1 in Chapter 5.1.
3. **False**. An allocation of given quantities of resources is efficient in production if it is not possible to reallocate the resources in such a way as to produce more of a good without producing less of another good. See Definition 5.1 in Chapter 5.1.
4. **True**. This is true by definition. See Definition 5.2 in Chapter 5.1.

5.1.2.2 Sample Solutions for Block 2

1. **False**. The First Theorem of Welfare Economics states that every equilibrium on competitive markets is Pareto-efficient and thus maximizes the sum of consumer and producer surplus. Only in one special case does this also imply this statement. See Result 5.1 in Chapter 5.2.
2. **True**. The allocation that maximizes the sum of consumer and producer surplus corresponds to a Pareto-efficient allocation. This is the case in the equilibrium in a polypoly. See Chapter 5.2.
3. **True**. Every Pareto optimum can be decentralized as a market equilibrium. See Result 5.2 in Chapter 5.2.
4. **False**. The First Theorem of Welfare Economics states that an equilibrium on a competitive market must be Pareto-efficient. However, it does not make any claims to exclusiveness, i.e. that every Pareto-efficient allocation must be an equilibrium on a competitive market. See Result 5.1 in Chapter 5.2.

5.1.2.3 Sample Solutions for Block 3

1. **False**. An allocation of given quantities of resources, goods, and services is Pareto-efficient if it is efficient in production and consumption. See Definition 5.3 in Chapter 5.1.
2. **False**. An allocation is called Pareto-efficient if it is efficient both in production and in consumption. However, one cannot conclude that it is efficient in consumption solely based on it being efficient in production. See Chapter 5.1.
3. **False**. An allocation is Pareto-efficient if it is both efficient in production as well as in consumption. One cannot determine that it is Pareto-efficient solely based on it being efficient in production. See Definition 5.3 in Chapter 5.1.
4. **True**. See Definition 5.3 in Chapter 5.1.

Externalities and the Limits of Markets

6

6.1 True or False

6.1.1 Statements

6.1.1.1 Block 1

A local government is thinking of prohibiting smoking in restaurants. Check the following arguments for their economic correctness. Assume that, by smoking, smokers have a negative interdependence with non-smokers.

1. A general smoking prohibition would lead to an efficient result, because it removes the potential external effect of smoking.
2. Smoky air is a public good, because neither the principle of rivalry nor the principle of excludability applies.

Now, assume that smokers and non-smokers negotiate in a restaurant and strike a deal. The smokers receive the right to smoke or the non-smokers receive the right for the smoking to cease.

3. Independent of the allocation of property rights, these negotiations lead to an optimal amount of smoked cigarettes in the restaurant if only a few customers are present.
4. The right to smoke leads to more smoked cigarettes than the right to fresh air.

6.1.1.2 Block 2

1. The originator of the external effect and the originator of the interdependency are one and the same.
2. Interdependencies are external effects that have not been internalized.
3. The Coase Irrelevance Theorem states that, in an economy with fully allocated property rights, the market equilibrium is always efficient.
4. If a group of individuals suffers from air pollution caused by a local chemical factory, this is a negative external effect.

M. Kolmar, M. Hoffmann, *Workbook for Principles of Microeconomics*,
Springer Texts in Business and Economics, https://doi.org/10.1007/978-3-319-62662-8_6

6.1.1.3 Block 3

1. Pollution is an externality.
2. Public goods and common goods differ with respect to their excludability from consumption.
3. Public roads are public goods.
4. Due to the fact that cinematic screening is excludable but non-rivalrous in consumption, it is a club good, as long as the show is not full.

6.1.1.4 Block 4

1. The radiation emitting from a nuclear power-plant after an accident is an interdependency for the affected habitants.
2. The insights gained from space travel are a public good for humankind.
3. The resource "tuna fish" is a private good.
4. The reach of an economic activity is the set of people directly influenced by the activity.

6.1.1.5 Block 5

1. If for a certain good the principle of excludability holds but the principle of rivalry does not, then it is a club good.
2. If for a certain good neither the principle of excludability nor the principle of rivalry holds, then it is a common good.
3. Radioactivity possesses all important characteristics of a club good.
4. Given the current level of technology, the surface of the moon is a common good.

6.1.1.6 Block 6

1. The Coase Irrelevance Theorem states that a market equilibrium is efficient because rational individuals will continue to negotiate until they have managed to extract all potential gains from trade.
2. The costs that arise while designing a contract are transaction costs.
3. Environmental externalities can be internalized through the utilization of legal liability laws.
4. The insights from the Coase Irrelevance Theorem are reflected in the "polluter-pays principle" and in liability laws.

6.1.1.7 Block 7

1. There is a market failure with public goods because there is no rivalry in consumption.
2. If fish populations migrate across exclusive fishing zones, it is to be expected that the populations are exploited in an inefficient manner.

3. If property rights are not, or not completely, enforceable, then there exist external effects on markets.
4. If public goods are sold in markets, then there are no external effects.

6.1.1.8 Block 8

1. If the production of a good causes positive externalities, then, according to the criterion of Pareto-efficiency, the good tends to be produced in too small quantities.
2. If a good is non-rivalrous in consumption, it cannot be supplied effectively using market mechanisms.
3. Competitive markets can also be efficient if there are positive or negative interdependencies.
4. The Coase Irrelevance Theorem implies that inefficient markets can either be traced back to positive transaction costs or missing property rights.

6.1.1.9 Block 9

1. Public goods should always be supplied by the government because it can produce them at lower costs than private companies.
2. It is impossible to exclude others from the consumption of club goods.
3. Public forests are a good example of public goods.
4. Oxygen is an example of a common good.

6.1.1.10 Block 10

1. If the two Theorems of Welfare Economics hold for all markets, it is possible for the government to guarantee Pareto efficiency by defining and enforcing property rights, as well as by enforcing contracts.
2. A night-watchman state is a state that restricts its activities to the enforcement of property rights and private contracts, as well as to the internalization of external effects.
3. External effects between generations can be internalized by means of contracts.
4. An external effect is a non-internalized interdependency.

6.1.1.11 Block 11

1. Good A has a negative externality in production and good B is complementary to good A. In this case, when compared to the Pareto-efficient allocation, too much of good B will be produced.
2. If the first Theorem of Welfare Economics holds true, then no externalities exist.
3. Market failure is understood as an institutional structure in which externalities occur in equilibrium.
4. The resources that are used for police and courts are part of the transaction costs of a market economy.

6.1.1.12 Block 12

1. A "market for lemons" describes a situation in which the seller knows pertinent information about the quality of a product that the buyer is missing and that raises the good's price.
2. If the seller has additional information about the quality of a product, that the buyer lacks, this can lead to a market in which only products of the highest quality are sold.
3. Externalities in traffic can be internalized through the means of road usage fees if transaction costs are not too high.
4. Liability law is an efficient means of internalizing externalities, that are caused by taking high risks, because it increases the perceived costs if the risk materializes.

6.1.1.13 Block 13

1. The cod fishery collapsed because the cod fish are migratory and thus a club good.
2. Markets fail when it comes to pollination services provided by bees because this is a public good (with limited reach).
3. If there are externalities, then it is economically efficient to make a polluter pay for the damages he or she caused.
4. The problems caused by externalities cannot be analyzed in a supply and demand diagram, because the problems occur due to the fact that a market for externalities cannot develop.

6.1.2 Solutions

6.1.2.1 Sample Solutions for Block 1

1. **False**. While the prohibition removes the externality, the result is not efficient. An efficient result can only be achieved by internalizing the negative interdependencies. See Chapter 6.2.
2. **True**. See the introductory remarks to Chapter 6.3.
3. **True**. The ownership rights (of the air) are fully allocated and, if there is a small number of guests, the transaction costs are close to zero. According to the Coase Irrelevance Theorem, the negotiations thus lead to an efficient result. See Chapter 6.2.3.
4. **False**. According to the Coase Irrelevance Theorem, the distribution of rights – as long as they are fully allocated – is irrelevant for the result. The negotiations will lead to the same quantity of smoked cigarettes, whether one has a right to smoke or a right to fresh air. See Chapter 6.2.3.

6.1.2.2 Sample Solutions for Block 2

1. **False**. There is no originator of an external effect. See Digression 12 in Chapter 6.2.
2. **False**. External effects are interdependencies that have not been internalized. See Definition 6.1 in Chapter 6.2.1.
3. **False**. The Coase Irrelevance Theorem states that in an economy with fully allocated property rights and without transaction costs, a market is always efficient. See Chapter 6.2.3.
4. **False**. It is only a negative externality if the interdependency is not internalized. See Definition 6.1 in Chapter 6.2.1.

6.1.2.3 Sample Solutions for Block 3

1. **False**. Externalities are interdependencies that have not been internalized. If the pollution's interdependencies could be completely internalized, then there would be no externalities. See Chapter 6.2.
2. **False**. Public goods and common goods differ as to whether they are rivalrous in consumption or not. See Chapter 6.3.
3. **False**. Especially during rush hour, there is rivalry in consumption (traffic jams). Therefore, many streets should be seen as common goods and not public goods. See Chapter 6.2.3.1.
4. **True**. This is true by definition. See Chapter 6.3.

6.1.2.4 Sample Solutions for Block 4

1. **True**. It is an interdependency, because the radiation influences the inhabitants' well-being. See Chapter 6.2.
2. **True**. They are neither excludable nor rivalrous, therefore they are public goods. See Chapter 6.3.
3. **False**. While one cannot exclude others from the resource tuna fish, it is rivalrous in consumption. Therefore, it is a common good. See Chapter 6.3.
4. **True**. This is true by definition. See Definition 6.3 in Chapter 6.3.

6.1.2.5 Sample Solutions for Block 5

1. **True**. This is true by definition. See Chapter 6.3.
2. **False**. If one cannot exclude others from its consumption and it is non-rivalrous in consumption, then it is a public good. See Chapter 6.3.
3. **False**. One cannot exclude others from consuming it and it is non-rivalrous in consumption. Therefore, radioactivity is a public good. See Chapter 6.3.
4. **True**. One cannot exclude others from its consumption and it is rivalrous in consumption. See Chapter 6.3. Note: One hundred years ago, the surface of the moon was a public good, as technology at that time did no allow for any alteration of the moon's surface and thus there was no rivalry in consumption.

6.1.2.6 Sample Solutions for Block 6

1. **False**. This is only the case if there are no transaction costs. There are two conditions for an efficient equilibrium: the involved parties' rationality and the institutions' lack of transaction costs. See Chapter 6.2.3.
2. **True**. Transaction costs due to formulating contracts: contracts do not just come into existence, they have to be negotiated. This requires time and specific expertise. See Chapter 6.2.3.
3. **False**. Liability law forces firms to pay for damages if they occur. Thus, the firms' costs in case of an accident increase, which would make it a theoretically promising instrument for internalizing externalities. However, this legal instrument can conflict with other legal instruments, which follow their own logic. Most countries, for example, have an insolvency law that restricts the risks of firms and individuals. If such a law is in place, liability law can become a toothless tiger, because the worst case scenario is insolvency, which decreases a firm's risks. For example, because oil spills generally carry high financial risks, liability law may be insufficient for causing the right incentives, because the owners are protected from high losses. See Chapter 6.2.3.2.
4. **False**. The "polluter-pays principle" assigns the financial responsibility for causing damage to a specific party. The Coase Irrelevance Theorem, however, states that there is no originator of an externality, i.e. there is no party that caused it. See Digression 12 in Chapter 6.2.2.

6.1.2.7 Sample Solutions for Block 7

1. **False**. Public goods can be consumed without needing to pay for them, because one cannot be excluded from consuming them. Thus, the fact that one cannot be excluded from its consumption leads to market failure. See Chapter 6.3.
2. **True**. In that case, the fish populations are common goods. Again, no one can be excluded from consuming them, meaning that no one will be willing to pay anything for them. Thus, externalities like exploitation occur, because the interdependencies cannot be internalized via a price. See Chapter 6.3.
3. **False**. Only if there are interdependencies that cannot be internalized due to a lack of property rights. See Chapter 6.2.
4. **False**. Interdependencies occur that cannot be internalized by means of a market. Thus, there are external effects. See Chapter 6.2.

6.1.2.8 Sample Solutions for Block 8

1. **True**. An allocation of given resources and quantities of goods is Pareto-efficient if it is both production and consumption efficient. If the production of a good causes positive externalities, then there are positive interdependencies that cannot be internalized. Thus, the supplier ignores these. If the interdependencies could be internalized, more of the good would be supplied. See Chapter 6.2.

2. **False**. If the good is non-rivalrous, then it is a club good or a public good. Club goods are perfectly suited to be provided using market mechanisms, for example, a movie theatre, music, etc. See Chapter 6.3.
3. **True**. If the interdependencies are internalized, then there are no externalities and the market is efficient. See Chapter 6.2.
4. **True**. The Coase Irrelevance Theorem states that, in an economy with perfectly allocated property rights and without transaction costs, the market equilibrium is always efficient. See Chapter 6.2.

6.1.2.9 Sample Solutions for Block 9

1. **False**. It is impossible to exclude others from the consumption of public goods. Thus, there is a market failure. If the government supplies public goods, being able to produce them at lower cost is not the reason. Rather, it is because of market failure due to non-excludability. See Chapter 6.3.
2. **False**. Exclusion is possible in club goods and there is no rivalry in consumption. See Chapter 6.3.
3. **False**. Public forests are common goods because there is rivalry in consumption but not excludability. See Chapter 6.3.
4. **True**. There is rivalry in consumption and it is impossible to exclude others from it. See Chapter 6.3.

6.1.2.10 Sample Solutions for Block 10

1. **True**. In this case, the role of the state is restricted to that of a night-watchman, a metaphor from libertarian political philosophy. See Chapter 6.1.
2. **False**. A night-watchman state's only legitimate purpose is the enforcement of property rights and contract law. See Chapter 6.1.
3. **False**. The interdependencies between generations cannot be internalized by means of contracts because one side of the market has not been born at the point of time when the contract is signed. See Chapter 6.2.3.
4. **True**. This is true by definition. See Definition 6.1 in Chapter 6.2.

6.1.2.11 Sample Solutions for Block 11

1. **True**. See Definition 4.6 in Chapter 4.2. See also Chapter 6.2.2.
2. **True**. See Result 5.1 in Chapter 5.2 (every equilibrium on competitive markets is Pareto-efficient). In perfect competition, every good is traded on a market and there exists a complete set of competitive markets (i.e. one for every interdependence). In order for externalities to exist, one must have uninternalized interdependencies. See Definition 6.1 in Chapter 6.2.
3. **True**. A situation in which externalities exist within a market system is also called a market failure. The concept of external effects refers to the institutional framework. See Chapter 6.2.1.

4. **True**. Transaction costs due to the enforcement of contracts: even in a night-watchman state, property rights and contractual arrangements have to be backed by the police and courts. See Chapter 6.2.3.

6.1.2.12 Sample Solutions for Block 12

1. **False**. Asymmetric information on part of the seller makes him unwilling to sell cars of higher quality in the market, because the higher quality would justify a higher price. Thus, the seller chooses not to supply high quality cars. Buyers realize this and adjust their expectations regarding the quality of the cars supplied in the market as well as their willingness to pay (i.e. expect only low quality cars (lemons) to be supplied and are thus, only willing to pay a low price). See Chapter 6.2.3.
2. **False**. See Chapter 6.2.3. Also see sample solution to Block 12, Statement 1.
3. **True**. The levying of congestion-dependent charges (for example during rush-hour traffic) can help solve the efficiency problem. See Chapter 6.2.3.1.
4. **False**. See sample solution to Block 6, Statement 3.

6.1.2.13 Sample Solutions for Block 13

1. **False**. Cod is a common good because exclusion is impossible and the good has a minimum reach. On the other hand, for club goods, exclusion is possible and the good is non-rivalrous in consumption. See Chapter 6.3.
2. **False**. The service provided by the bees is not a public good. First of all, the services are rivalrous in consumption (the bee can pollinate one or the other fruit tree). Second, the person selling the pollination service can offer it to either orchardist 1 or 2; thus, it is excludable. Therefore, it is a private good. See Digression 13 in Chapter 6.2.2.
3. **False**. See Digression 12 in Chapter 6.2.2. Also have a look at the Coase Irrelevance Theorem in Chapter 6.2.
4. **False**. Example: Emissions (the production of bread generates sewage). In this case the markets are incomplete, since there is no market for emissions. If we want to analyze the problem in the context of demand and supply, we have to look at a market that exists. In this case, this is the market for bread. See Chapter 6.2.2.

6.2 Open Questions

6.2.1 Problems

6.2.1.1 Exercise 1

You are the advisor to a city that is suffering from local emissions (regional effects), and must develop a program to solve the issue.

Table 6.1 Exercise 2. Guests and profits on *Silentium* and *Gaudium*

Silentium		*Gaudium*	
Guests	Profit	Guests	Profit
0	0	0	0
10	10*k*	100	50*k*
20	12*k*	200	60*k*
30	13*k*	300	65*k*
40	11*k*	400	66*k*

1. Explain to the city's mayor how the terms *interdependencies*, *externalities* and *internalization* are linked to each other.
2. Discuss if the emissions are causing efficiency problems for the city.
3. Discuss how likely it is that efficiency would be attainable by distributing ownership rights.
4. Discuss if a law, that defines emission caps of the individual firms, would lead to efficient levels of emission.
5. Assume that the city has capped the total emissions for all firms, and now offers emission rights to the size of the optimal total emission. Thus, the firms can only produce the amount of emissions for which they have purchased the ownership rights. How does this differ from the market mechanism that is discussed in Question 3?
6. How does the argumentation above change if the emissions have not only local but global effects?

6.2.1.2 Exercise 2

There are two neighboring vacation islands that differ by their target groups. Each of these islands is managed by a different hotel chain. The island *Silentium* mainly draws customers who are looking for peace and relaxation. Opposite them is the island *Gaudium*, which mainly attracts students who like to have long, loud parties. Because of the small distance between the islands, the hotel on *Silentium* (also called *Silentium*) has to offer a price reduction to its customers if they feel disturbed by the noise emissions from the hotel on *Gaudium* (also called *Gaudium*). To simplify we assume that the profit of *Silentium* drops by 2,000 Swiss Francs per hundred guests on *Gaudium*. The gains of both islands, which would result if there were no interdependencies between the two islands, are illustrated in Table 6.1.

1. Is there an externality? If yes, which hotel causes it?
2. Illustrate in a suitable table, how the profits of *Silentium* develop in correspondence with the amount of guests on *Gaudium*.
3. Which combination of guests would maximize total profits for both hotels? Present the allocation of these guests and the profits for both of the hotels.
4. Imagine that *Silentium* has the necessary property rights to demand the retribution of the loss of profit caused by the noise emissions from *Gaudium*. How many guests would both hotels welcome and how high are the profits of the hotels? How does this differ from the result in Question 3?

5. How would the situation change if *Gaudium* has the necessary property rights so that it does not have to pay for any reduction of profits of *Silentium*? Compare your result with the results found in Questions 3 and 4.

6.2.1.3 Exercise 3

After winning in the lottery, Igor decides to found his own country on a distant uninhabited island in the Pacific that he bought with the money. He wants to put on some fireworks to celebrate the foundation. This service's inverse supply function is $Q(y) = 3$, in which y is the firework's length in minutes. Igor's demand function for the fireworks (also in minutes) is $x^I(p) = 20 - p$.

1. Assume that Igor lives on the island alone. How high is Igor's marginal willingness to pay corresponding to the firework's length? Determine the firework's Pareto-efficient duration.
2. By chance, Igor's great aunt Gertrude finds out about him winning the lottery and purchasing the island. She decides to visit her great nephew for the foundation celebration. Gertrude's demand function for the firework is $x^G(p) = 20 - 2p$. Determine the firework's Pareto-efficient duration and the consumer surplus (CS) in this case. Compare this result to that of Question 1. Is it possible to derive a rule for the Pareto optimal fireworks supply?
3. Igor decides to cover the costs for supplying the fireworks himself. What is the consumer surplus (CS) in this case? Explain the result.
4. Igor decides to make his great aunt cover some of the firework's cost. Assume that there are the following *individualized* prices for a unit x: $p^G = \tau\, p$ for Gertrude and $p^I = (1 - \tau)\, p$ for Igor, in which $0 \leq \tau \leq 1$.
   a) Assume that $\tau = \frac{1}{2}$. What are the quantities the two of them demand now? What is noteworthy?
   b) Assume that $\tau = \frac{1}{3}$. What are the quantities the two of them demand now? What is noteworthy?

6.2.1.4 Exercise 4

The company *Peter's Paper Mill* produces paper with the aid of river water. This produces waste water in the amount of x cubic meters, which are discharged into a nearby lake. The value of x at full capacity of the company is $\bar{X} = 1{,}000$. The profit of *Peter's Paper Mill* (at full capacity) depending on the waste water pumped into the lake is

$$\Pi_P(x) = 2{,}100\, x - x^2$$

and is illustrated in Fig. 6.1.

The marginal profit of the waste water dumping is $\Pi'_P(x) = 2{,}100 - 2x$ (remember that this represents opportunity costs).

1. Discharging the waste water into the lake impairs an adjacent fisherman. His marginal profit from the polluted lake is $\Pi'_F(y) = 50 - \frac{1}{20}\, y$, with $y = \bar{X} - x$.
   a) Illustrate the marginal profit of the company through the discharge of waste water ($\Pi'_P(x)$) in a diagram. How high is the profit at $x = \bar{X}$?

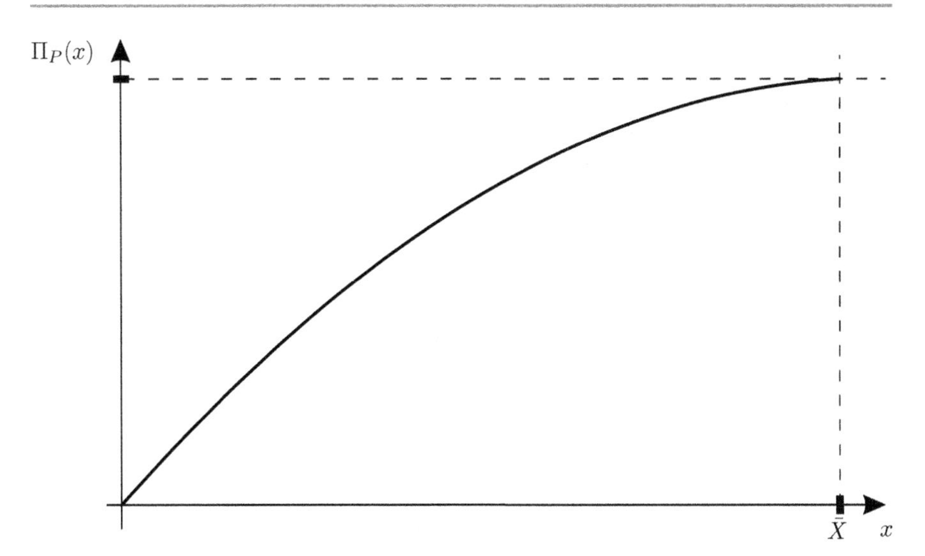

Figure 6.1 Exercise 4. Profit of *Peter's Paper Mill* depending on the quantity of waste water dumping x

b) Illustrate the marginal profit of the fisherman through avoiding the discharge of waste water ($\Pi'_F(y)$) in the same diagram. How high is the profit ($\Pi_F(y)$) at $y = \bar{X}$?

c) Assume that the company owns the right to discharge the waste water in the lake. What would the market solution look like in that case? How high are the compensation costs of the fisherman? How high is the total profit?

d) Assume that the fisherman owns the rights to the lake. What would the market solution look like in that case? How high are the compensation costs of the company? How high is the total profit?

2. The property rights of the lake belong to *Peter's Paper Mill.* Assume that there is not one fisherman, but 8 fishermen with identical marginal profit functions $\Pi'_{F_i}(y) = 50 - \frac{1}{20}y$, $i \in \{1, \ldots, 8\}$. Furthermore, let us assume that fixed transaction costs of $c \geq 0$ *per fisherman* are caused during a negotiation.
   a) Let $c = 0$. What would be the market solution in that case? Illustrate the net-profit of both of them (the profit minus or plus the possible compensation costs) in a diagram. Illustrate the total profit as well.
   b) What level of the transaction costs lead to an inefficient market solution?
   c) If the transaction costs are larger than zero but smaller than the ones calculated in Question 2b), does this contradict the Coase Irrelevance Theorem?

6.2.2 Solutions

6.2.2.1 Solutions to Exercise 1

1. *Interdependencies* always show up wherever decision makers influence the sphere of other individuals through their actions. If all interdependencies are *internalized* by means of competitive prices, the market result is efficient. For example, the employer causes a opportunity cost on the employee, which is compensated by a salary. However, if there are some whose *interdependencies* are not *internalized*, the market equilibrium is inefficient. The non-internalized interdependencies are called *externalities*.
2. The information given is insufficient to reach a conclusion on the possible inefficiencies. Depending on the circumstances, all of the interdependencies may have already been internalized.
3. If the ownership rights are completely distributed and enforced, then either the citizens have a right to clean air or the firms a right to cause pollution. This is, according to Coase, a necessary condition for an efficient allocation of the property rights by means of market prices. The second condition is that the transaction costs have to be zero. If this is true, then the market result is efficient. However, the absence of transaction costs is a theoretical abstraction, which means that in practice, you have to assess how high the transaction costs are for a specific institution, compare them, and choose the institution that minimizes them. In our specific example, we can assume that decentralized negotiations between the inhabitants and the firms would cause such transaction costs of reaching an agreement and that this kind of a solution would not work.
4. If the planners of the city have complete information and know exactly how high the ideal emission for each industrial firm would be, they could create an efficient level of emissions. For varying cost-structures, one can assume that the costs of avoiding emissions is different for each firm as well and thus provides the optimal prevention of emissions. However, this knowledge is generally not available. This means that, for example, consistent emission standards lead to inefficiencies, because the reduction of emission costs does not minimize costs. Depending on the transaction costs of the other solutions, this solution may be viable or not. In comparison to the decentralized negotiations from Question 3 this would be an improvement.
5. If the total quantity of emissions is set at an optimal level, and thus corresponds to the efficient amount of pollution, then the efficient distribution of the emissions can be set by the efficient arrangement of the market for emission certificates. The trading of emission rights will cause the minimal costs to be reached for a specific goal for emission if the market is well-constructed. Possible issues are caused by the setting of an efficient goal for emission and the design of the market, especially if there are very few firms emitting a specific pollutant. Every firm will continue to buy emission certificates until the price of an additional certificate corresponds to the margin of pollution-avoidance costs.

Table 6.2 Exercise 2.2. Profits on *Silentium* depending on the guests on *Silentium* and *Gaudium*

	Guests *Gaudium*				
Guests *Silentium*	0	100	200	300	400
0	0	0	0	0	0
10	$10k$	$8k$	$6k$	$4k$	$2k$
20	$12k$	$10k$	$8k$	$6k$	$4k$
30	$13k$	$11k$	$9k$	$7k$	$5k$
40	$11k$	$9k$	$7k$	$5k$	$3k$

Thus, the price for a certificate will even out so that the margin of avoidance-costs is identical for all the firms. Thus, it would be efficient.
In comparison to Question 3 the population was automatically given a right to clean air. In Question 3 the firms could have also been granted the right to pollute. Additionally, in this example, the goal of reaching the efficient allocation in the market for pollution rights was implemented by an institution.

6. The argument remains the same. However, all firms emitting in the various countries have to be included as well now. Since these emissions take place across borders, a supra-institution would have to be created, that would control the commerce and define the efficient total amounts.

6.2.2.2 Solutions to Exercise 2

1. Initially one would have to clarify if noise pollution is an externality. If the interdependency caused by the guests' behavior is internalized by suitable prices, then externalities would not even emerge. However, even if these internalizations did not exist, the question of who caused the externalities would be pointless. For instance, if there were no guests on *Silentium*, there would be no interdependency and the noise emission on *Gaudium* would not bother anyone. So, to claim that *Gaudium* is the cause of the externality, due to their noise emissions is just as legitimate (or illegitimate) as the claim that the guests on *Silentium* are causing the externality.
2. The profits of the hotel on *Silentium* depending on the number of guests on *Silentium* and *Gaudium* is illustrated by Table 6.2.
3. If we want to maximize total profits, we have to compare and check all possible numbers of guests on both islands for their efficiency. Drawing on the table from Question 2 (Table 6.2), we can find five different combinations from which to choose (see Table 6.3). Take into consideration that *Silentium* always welcomes 30 guests to maximize its profits.
It is easy to tell that total profits are at the maximum at 72,000 Swiss Francs. This means that, to maximize total profits, you would place 300 guests on *Gaudium* and 30 on *Silentium*. This leads to a profit of 65,000 Swiss Francs on *Gaudium* and 7,000 Swiss Francs on *Silentium*.
4. If the island *Silentium* owns the property rights to demand retribution of profit reductions caused by the noise emissions from *Gaudium*, then *Silentium* will

Table 6.3 Exercise 2.3. Determination of the optimal total profit

Gaudium		*Silentium*		
Guests	Profit	Optimal guests	Profit	Total profit
0	0	30	13*k*	13*k*
100	50*k*	30	11*k*	61*k*
200	60*k*	30	9*k*	69*k*
300	65*k*	30	7*k*	72*k*
400	66*k*	30	5*k*	71*k*

Table 6.4 Exercise 2.4. Optimization plan for *Gaudium*

Guests *Gaudium*	Profit – Compensation	Residual Profit
0	0	0
100	50*k* – 2*k*	48*k*
200	60*k* – 4*k*	56*k*
300	65*k* – 6*k*	59*k*
400	66*k* – 8*k*	58*k*

choose the number of guests as if there were no interdependencies between the two islands. Thus, *Silentium* will welcome 30 guests and will earn 13,000 Swiss Francs. *Gaudium* has to bear in mind that its profits per 100 guests are lowered by 2,000 Swiss Francs through the compensation payments to *Silentium* and will choose an optimal number of guests. The possibilities as illustrated in Table 6.4 ensue.

Gaudium maximizes its profits if it welcomes 300 guests. The number of guests thus remains the same as in Question 3. The difference is the distribution of profits. If the ownership rights belong to *Silentium*, it has profits of 13,000 Swiss Francs while *Gaudium* has profits of only 59,000 Swiss Francs now.

5. If *Gaudium* owns the property rights, and thus does not have to pay for any profit losses caused by the noise emissions, they will want to maximize their profit by welcoming 400 guests (see Table 6.1). We already know that the profits of *Silentium* is at the maximum at 30 guests, regardless of how many guests are on *Gaudium*; thus, *Silentium* is making a profit of 5,000 (see Table 6.2). However, the island *Silentium* can also negotiate with *Gaudium*, so they receive fewer guests. In compensation they have to pay *Gaudium* the amount of the foregone profit. *Silentium* has the following options (see Table 6.5).

   Silentium will compensate *Gaudium* with 1,000 Swiss Francs so that they only allow 300 guests on their island. This increases *Silentium*'s profit by 1,000 to 6,000 Swiss Francs in comparison to the initial situation (with 400 guests on *Gaudium*). Thus, the allocation of guests is identical in this situation to the ones in Questions 3 and 4. The only thing that changed, once again, is the distribution of profits.

   In this exercise it was implicitly assumed that the transaction costs are zero and the ownership rights are well defined and perfectly enforced. In Question 3, the efficient solution was defined, and in Questions 4 and 5 we found a decentralized

Table 6.5 Exercise 2.5. Optimization for *Silentium*

Gaudium		*Silentium*			
Guests	Profit	Guests	Profit	Compensation	Net-profit
0	0	30	$13k$	$66k - 0 = 66k$	$13k - 66k = -53k$
100	$50k$	30	$11k$	$66k - 50k = 16k$	$11k - 16k = -5k$
200	$60k$	30	$9k$	$66k - 60k = 6k$	$9k - 6k = 3k$
300	$65k$	30	$7k$	$66k - 65k = 1k$	$7k - 1k = 6k$
400	$66k$	30	$5k$	$66k - 66k = 0$	$5k - 0 = 5k$

Table 6.6 Exercise 2.5. Total profits

Question	Guests		Profit		Total Profit
	Silentium	*Gaudium*	*Silentium*	*Gaudium*	
3	30	300	$7k$	$65k$	$72k$
4	30	300	$13k$	$59k$	$72k$
5	30	300	$6k$	$66k$	$72k$

solution, where we assumed that either *Gaudium* or *Silentium* has a right to noise or a right to avoid noise. As we can see, all the results regarding the allocation of guests are identical, since there were always 30 guests on *Silentium* and 300 guests on *Gaudium* (see Table 6.6).
Total profits of both islands are also identical. Thus, this problem also confirms the Coase Irrelevance Theorem, which says that under the assumption of perfectly defined property rights and without transaction costs the result of the negotiations between the parties is efficient (see Chapter 6.2.3). The distributional results, on the other hand, are not identical, because they depend on which party owns the property rights (see Table 6.6).

6.2.2.3 Solutions to Exercise 3

1. The marginal willingness to pay for x is represented by the inverse demand function. This results in:

$$x^I(p) = 20 - p$$
$$\Leftrightarrow \quad P^I(x) = 20 - x.$$

According to Definition 5.3 in Chapter 5.1, an allocation of given resources and quantities of goods is Pareto efficient if it is both production and consumption efficient, i.e. if it maximizes the sum of producer and consumer surplus. In our case, the producer surplus is always zero, because the supply function is infinitely elastic ($\varepsilon_p^y \to \infty$, see Chapter 14.3). The consumer surplus corresponds to the area underneath the demand function $x^I(p)$ and above the price, $p = Q(y) = 3$. This area is maximized at the point where $P^I(x) = p$, i.e.,

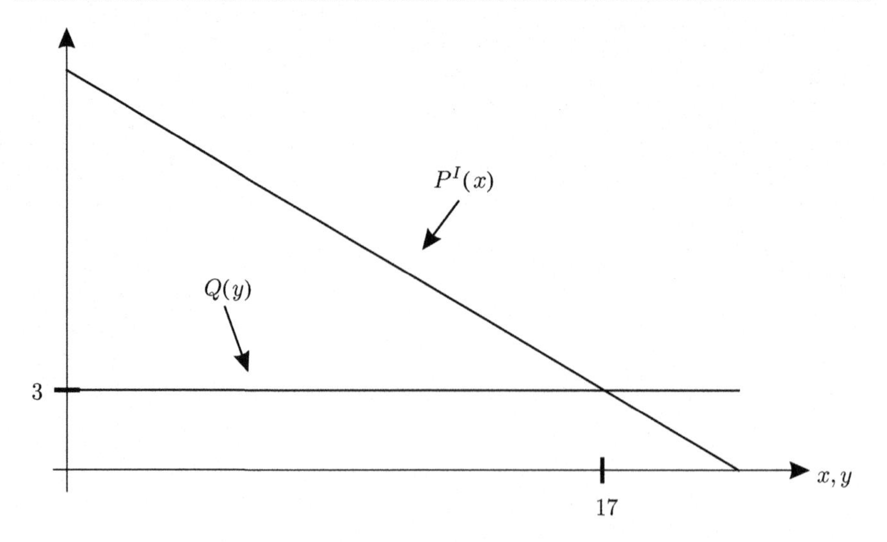

Figure 6.2 Exercise 3.1. Demand and supply function

where

$$\begin{aligned} 20 - x &= 3 \\ \Leftrightarrow \quad x^* &= 17. \end{aligned}$$

Thus, in the Pareto optimum, the firework will last for 17 minutes (see Fig. 6.2).

2. Fireworks are local public goods (see Chapter 6.3). Public goods are non-rivalrous in consumption and non-excludable. The fact that it is non-rivalrous has an influence on the Pareto-efficient quantity x. Every minute of the fireworks can thus be enjoyed by both Igor and his great aunt. The other person's presence, however, does not decrease how enjoyable the fireworks are. This also has an effect on the marginal willingness to pay, which is needed for calculating the Pareto optimum (see Question 1): The fact that another person is watching the fireworks does not have any influence on the individual's marginal willingness to pay. The total marginal willingness to pay for x, thus, corresponds to Igor's and Gertrude's aggregate willingness to pay.[1]
In order to determine this sum, one needs to determine Gertrude's inverse demand function. The aggregate marginal willingness to pay corresponds to the sum of the inverse demand functions. Gertrude's inverse demand function is:

$$\begin{aligned} x^G(p) &= 20 - 2\,p \\ \Leftrightarrow \quad P^G(x) &= 10 - 0.5\,x. \end{aligned}$$

[1] Think about it this way: what would change if Igor were to decide to serve champagne, a private good, instead of fireworks? Every glass that Igor drinks cannot be consumed by Gertrude. Thus, Gertrude's willingness to pay for that specific glass of champagne is zero and the aggregate willingness to pay for that specific glass of champagne equals Igor's willingness to pay.

Then, the aggregate willingness to pay is:

$$\begin{aligned} & P(x) = P^I(x) + P^G(x) \\ \Leftrightarrow \quad & P(x) = 20 - x + 10 - 0.5\,x \\ \Leftrightarrow \quad & P(x) = 30 - 1.5\,x. \end{aligned}$$

One uses an analogous approach to Question 1 to determine the Pareto optimum. Consumer surplus is given by the area between the inverse demand function and the price line. This area is maximized at $P(x) = p$, so that

$$\begin{aligned} & 30 - 1.5\,x = 3 \\ \Leftrightarrow \quad & x^* = 18. \end{aligned}$$

This result is illustrated in Fig. 6.3.
Hence, in the Pareto optimum, the fireworks will last for 18 minutes. The presence of an additional person (Gertrude), thus, increases the quantity provided in the Pareto optimum by one minute. Because $P(x)$ is a linear function, one can use the triangle-formula to calculate consumer surplus at x^*[2]

$$CS(x^*) = (P(0) - p) \cdot x^* \cdot \frac{1}{2} = (30 - 3) \cdot 18 \cdot \frac{1}{2} = 243,$$

which is represented by the gray area in Fig. 6.3. Apparently, in the Pareto optimum, the sum of the individuals' marginal willingness to pay, at the point x^*, corresponds to the costs for supplying the fireworks (which are represented by the market price $p = 3$ in this example):

$$\begin{aligned} & P^I(x^*) + P^G(x^*) = p \\ \Leftrightarrow \quad & \underbrace{20 - x^*}_{=P^I(x^*)} + \underbrace{10 - 0.5\,x^*}_{=P^G(x^*)} = p \\ \Leftrightarrow \quad & 20 - 18 + 10 - 9 = 3. \end{aligned}$$

As long as $P(x) > p$, the quantity supplied should be extended in order to achieve a Pareto optimum. Then the costs for the last marginal unit (i.e. at the point x^*) will correspond to the sum of both individuals' marginal willingness to pay.

[2] Of course, using the general formula (see Definition 5.4 in Chapter 5.2) yields the same result for the consumer surplus:

$$CS(x^*) = \int_{x=0}^{x=x^*} (P(x) - p)d\,x = \int_{x=0}^{x=18} (27 - 1.5\,x)d\,x = \left[27\,x - 0.75\,x^2\right]_{x=0}^{x=18} = 243.$$

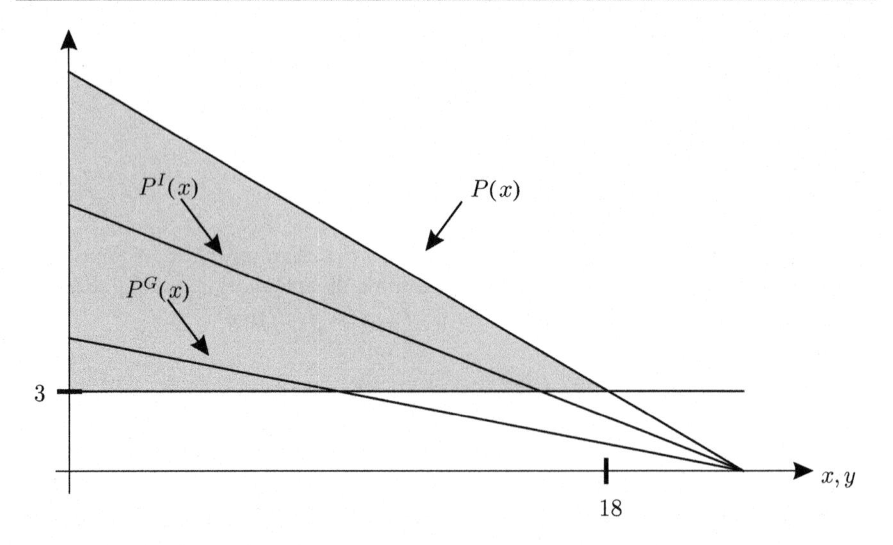

Figure 6.3 Exercise 3.2. Consumer surplus in the Pareto optimum

3. Because Igor wants to cover the costs for the fireworks himself, he will orient himself according to his own demand function. Thus, the following quantity will be supplied:
$$x^I(p = 3) = 20 - 3 = 17.$$
Igor ignores the positive interdependency: Gertrude has a positive willingness to pay for every minute of the fireworks (as long as $x < 20$), which Igor's calculations ignore. Thus, there is no internalization of the interdependencies, and a positive consumption externality emerges (see Chapter 6.2.2). The way a positive externality expresses itself is by supplying too little of a good in comparison to the Pareto optimum, $x^I < x^*$. Thus, the consumer surplus decreases in comparison to the Pareto optimum, because the following amount is lost:
$$\Delta CS = (P(17) - p) \cdot (x^* - x^I) \cdot \frac{1}{2} = (4.5 - 3) \cdot (18 - 17) \cdot \frac{1}{2} = 0.75.$$
Thus, in this case, the CS is:
$$CS(x^I) = CS(x^*) - \Delta CS = 243 - 0.75 = 242.25,$$
which is represented by the gray area in Fig. 6.4.
4. a) Now, Igor is requesting that his great aunt pay for half of the costs of the fireworks. Gertrude's individual price for x is thus $p^G = 0.5 \cdot 3 = 1.5$. This is also Igor's individual price ($p^I = 0.5 \cdot 3 = 1.5$). At these prices, they

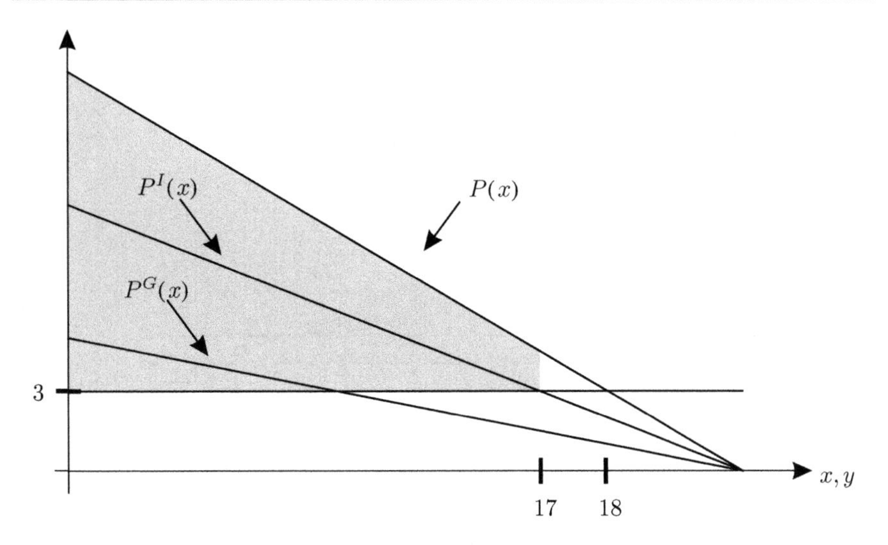

Figure 6.4 Exercise 3.3. Consumer surplus

demand the following quantities:

$$x^I(p^I = 1.5) = 20 - p^I = 20 - 1.5 = 18.5,$$
$$x^G(p^G = 1.5) = 20 - 2p^G = 20 - 2 \cdot 1.5 = 17.$$

In this case, Igor is trying to internalize the interdependencies by asking Gertrude to help pay for the fireworks. However, he has overstepped the mark, because Gertrude's demand is now smaller and his own is larger than the Pareto optimum would be ($x^G(p^G = 1.5) < x^* < x^I(p^I = 1.5)$).

b) The following quantities are demanded (Fig. 6.5):

$$x^I(p^I = 2) = 20 - p^I = 20 - 2 = 18,$$
$$x^G(p^G = 1) = 20 - 2p^G = 20 - 2 \cdot 1 = 18.$$

Apparently, this distribution of costs ensures a Pareto-efficient supply of the public good, because $x^I(p^I = 2) = x^G(p^G = 1) = x^* = 18$. Looking at the personalized prices more carefully shows that the following holds:

$$p^I = P^I(x^*)$$
$$\Leftrightarrow \quad 2 = 20 - x^* = 20 - 18$$
$$\text{and}$$
$$p^G = P^G(x^*)$$
$$\Leftrightarrow \quad 1 = 10 - 0.5\,x^* = 10 - 9.$$

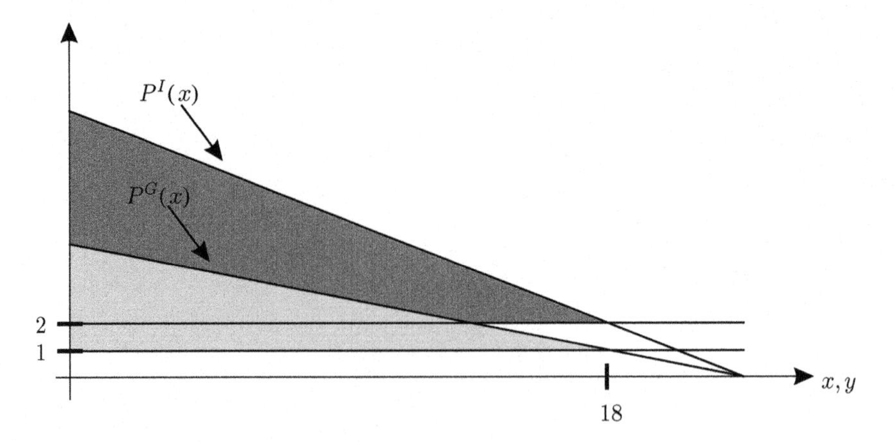

Figure 6.5 Exercise 3.4b). Individual demand functions and individual surpluses

Apparently, there is a market mechanism that guarantees a public good's Pareto-efficient supply: the individualized prices (p^I and p^G) have to be set such that the individual demand for the public good is identical ($x^I(p^I) = x^G(p^G)$). Then the individualized prices correspond to the individuals' marginal willingness to pay at the point $x^I = x^G$. Unfortunately, this method has a small flaw: it is necessary to know everyone's willingness to pay.

Why would this be necessary? Assume the following: Gertrude claims to have a willingness to pay of zero. In that case, the only relevant demand for the Pareto optimal supply of the public good is Igor's and his individualized price is $p^G = 3$. That result is already known from Question 3: the fireworks will last for 17 minutes. That is obviously not the Pareto optimum, because Gertrude's true willingness to pay is $P^G(x) = 10{-}2\,x$. However, Gertrude's false disclosure of her willingness to pay was worthwhile for her: she can watch a 17 minute fireworks show without paying for it. Thus, she acts as a free rider (the same argument can be made for Igor in the opposite case).

6.2.2.4 Solutions to Exercise 4

1. In this part of the exercise you should only focus on the interests of one fisherman and the company.
   a) The profit from disposing the waste water corresponds to the integral of the marginal profit and was already given in the problem definition. At $x = \bar{X}$, the profit corresponds to

   $$\Pi_P(\bar{X}) = 2{,}100\,\bar{X} - \bar{X}^2 = 1.1 \cdot 10^6.$$

   This amount is represented by the area underneath the marginal-profit function of the company (Π_P') in Fig. 6.6.

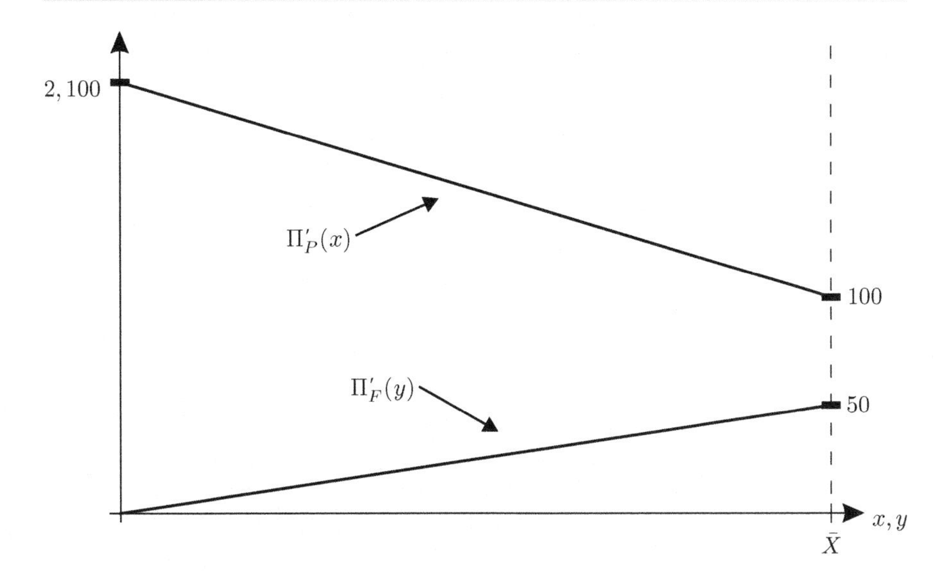

Figure 6.6 Exercises 1.1a) and 1.1b). Marginal profit of *Peter's Paper Mill* as a function of x ($\Pi'_P(x)$) and the fisherman's marginal profit as a function of y ($\Pi'_F(y)$)

b) The profit of the fisherman from the unpolluted lake (if $y = \bar{X}$) corresponds to the integral of the marginal profit $\Pi'_F(y)$.

$$\Pi_F(\bar{X}) = \int_0^{\bar{X}} \Pi'_F(y)d\,y = \left[50\,y - \frac{1}{40}\,y^2\right]_0^{\bar{X}} = 50\,\bar{X} - \frac{1}{40}\bar{X}^2 = 25{,}000.$$

This amount is represented by the area underneath the marginal-profit function of the fisherman in Fig. 6.6.

c) In this case, the marginal willingness to sell of the company (Π'_P) is always higher than the marginal willingness to pay of the fisherman (Π'_F, see Fig. 6.6). Thus, there are no gains from trade that can be exhausted and the company superimposes $x = \bar{X}$ (and thus $y = 0$). Hence, $\Pi_P = \Pi_P(\bar{X}) = 1.1 \cdot 10^6$ and $\Pi_F = \Pi_F(0) = 0$. The total profit is

$$\Pi = \Pi_P + \Pi_F = 1.1 \cdot 10^6.$$

d) In this case, the marginal willingness to sell of the fisherman (Π'_F) is always lower than the marginal willingness to pay of the company (Π'_P, see Fig. 6.6). The gains from trade will be exhausted when the fisherman sells all his property rights to the company. Principally, the fisherman can demand any price for this that is lower than the profit of the company at $x = \bar{X}$ ($\Pi_P(\bar{X})$). We assume here, though, that both sides of the market act as a price takers (thus as quantity-adjusters), and that the price per cubic meter of polluted water corresponds to the lowest marginal willingness to pay of the company. Thus, we obtain the price of p per x:

$$p = \Pi'_P(\bar{X}) = 100.$$

Thus, the price of the compensation of the company for the fisherman is:

$$K_P = p \cdot \bar{X} = 100{,}000.$$

The loss of profit the fisherman faces from the complete sale of the property rights to the company is $\Pi_F(\bar{X})$ and is represented by the area C in Fig. 6.7. The area $B+C$, on the other hand, corresponds to the compensation payment by the company, so the area B illustrates the profit of the fisherman from the sale. The profit of the company due to the purchase of all the rights corresponds to $\Pi_P(\bar{X})$ and is represented by the areas $A + B + C$. The net-profit (the profit from the purchase of all the rights after the deduction of the compensation payment) on the other hand only corresponds to the area A.

The profits of the two parties are $\Pi_P = \Pi_P(\bar{X}) - K_P = 1.1 \cdot 10^6 - 10^5 = 10^6$ and $\Pi_F = \Pi_F(0) + K_P = 10^5$ and the total profit is

$$\Pi = \Pi_P + \Pi_F = 1.1 \cdot 10^6,$$

which corresponds to the sum of the areas $A + B + C$ in Fig. 6.7.

This part of the exercise should illustrate that an efficient negotiated settlement can also mean that in the end, all or no property rights will be sold. While there were no feasible gains from trade in Question 1c), they were exhausted in Question 1d) through the complete sale of property rights. The sole fact that the two (potential) contractual parties do not start trading is no indication for the efficiency of the result of this trade. This will be clarified further in the next problem section.

2. The aggregated marginal-profit function of all fishermen is as follows:

$$\Pi'_F(y) = \sum_{i=1}^{8} \Pi'_{F_i}(y) = 400 - 0.4\,y.$$

a) In this case, the marginal willingness to pay of the fishermen is above the marginal willingness to sell of the company as soon as $y < y^*$. We define y^* as the intersection of both marginal-profit functions (see Fig. 6.8). For this purpose, it is necessary to define the marginal-profit function of the company via y instead of x. Because $x = \bar{X} - y$, it follows that

$$\Pi'_P(y) = 2{,}100 - 2(\bar{X} - y) = 100 + 2\,y.$$

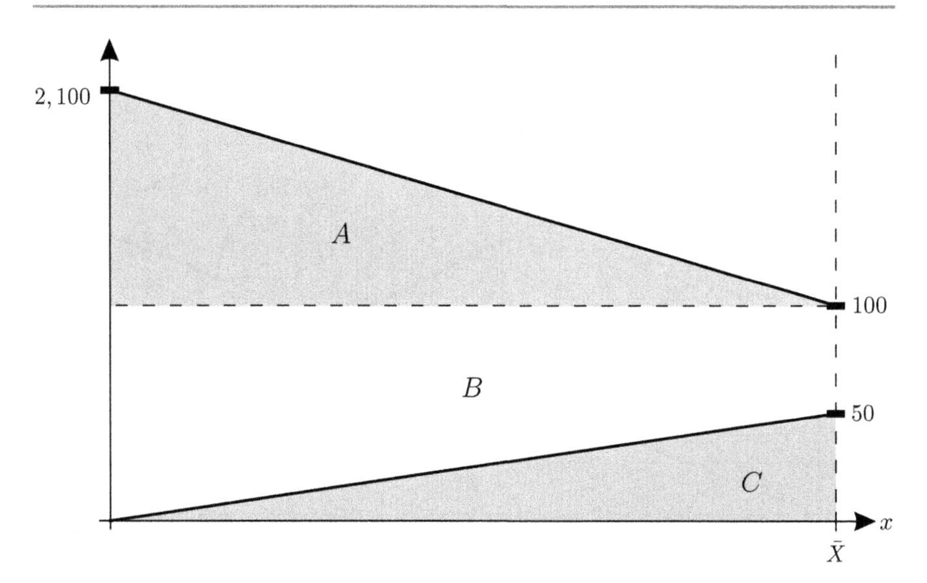

Figure 6.7 Exercises 1.1c) and 1.1d). Total profit

Thus, it follows that at the intersection of both marginal-profit functions:

$$\begin{aligned} \Pi'_F(y) &= \Pi'_P(y) \\ \Leftrightarrow \quad 400 - 0.4\,y &= 100 + 2\,y \\ y^* &= 125. \end{aligned}$$

Hence, the fishermen buy the property rights of 125 cubic meters of water. If we assume that the price (p) per cubic meter of water corresponds to the fishermen's marginal willingness to pay at the point $y = y^*$, we get:

$$p = \Pi'_F(y^*) = 400 - 0.4 \cdot y^* = 400 - 0.4 \cdot 125 = 350.$$

Thus, the compensation costs paid by the fishermen are as follows:

$$K_F = p \cdot y^* = 43{,}750.$$

The compensation payment is highlighted by the gray shaded area in Fig. 6.8 (area $E + F$). Through the reduction of x by 125 units, the company only relinquishes the area F, and is thus better off through trade. The profit of the company consists of the received compensation payment (area $E + F$) and the profit from the disposal of $\bar{X} - y^* = 875$ cubic meters of waste water, $\Pi_p(\bar{X} - y^*) = 1{,}071{,}875$ (area $A + B + C$):

$$\Pi_p = \Pi_p(\bar{X} - y^*) + K_p = 1{,}115{,}625.$$

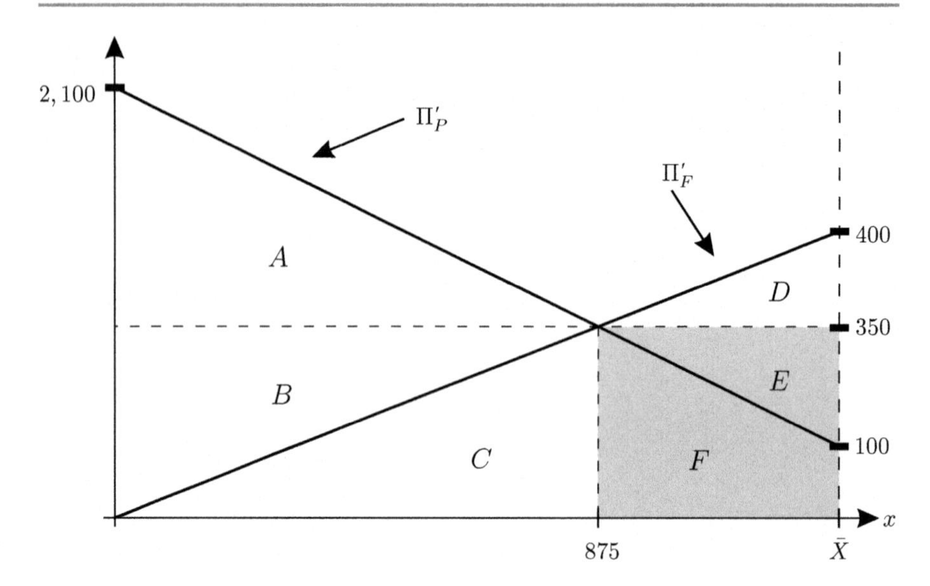

Figure 6.8 Exercise 1.2. Compensation payment

The profit of the fishermen from the reduction of the disposal of waste water by 125 units corresponds to the area underneath the marginal-profit function Π'_F (area $D + E + F$). The compensation payments have to be deducted from this (area $E + F$), so the area D illustrates the net-profit of all fishermen after the trade,

$$\Pi_F = \Pi_F(y^*) - K_F = (400 - 350) \cdot 125 \cdot \frac{1}{2} = 3{,}125.$$

If we assume that the compensation payments to the company are covered in equal parts by all fishermen ($K_F = 8 \cdot K_{F_i}$), the corresponding individual net-profit of every fisherman (after trading) is

$$\Pi_{F_i} = \Pi_{F_i}(y^*) - K_{F_i} = \frac{3{,}125}{8} = 390\frac{5}{8}.$$

This result is efficient, since transaction costs were assumed to be zero. Total net profit in the case of efficiency thus is

$$\Pi = \Pi_P + \Pi_F = 1{,}115{,}625 + 3{,}125 = 1{,}118{,}750,$$

which corresponds to the area $A + B + C + D + E + F$ in Fig. 6.8.

b) Because the transaction costs are fixed costs, and thus are independent from the traded amount x (or y), they only have an influence on the post-trade allocation of property rights if these transaction costs are higher than the net-profit of each individual fisherman, i.e. if

$$c > 390\frac{5}{8}.$$

In that case, the fishermen would prefer not to start negotiating with the company. Then, the sum of the transaction costs would be larger than the area D and total profit corresponds to the profit of the company, i.e. to the area $A + B + C + F$ in Fig. 6.8. Hence, it is much smaller than in Question 2a).

c) In this case, the transaction costs are strictly positive but smaller than the value at which the net-profits of the fishermen are over-compensated by the transaction costs. This means that the allocative result corresponds to the efficient result from Question 2a), i.e. $x = \bar{X} - y^* = 875$. This does not contradict the Coase Irrelevance Theorem because the latter only states that, if property rights are clearly defined and enforced and in the absence of transaction-costs, the market result must be efficient. However, the reverse does not hold. So, if one of the assumptions is not fulfilled, the market result does not necessarily have to be inefficient.

6.3 Multiple Choice

6.3.1 Problems

6.3.1.1 Exercise 1

A profit-maximizing firm acts as price-taker and produces according to the following inverse supply function: $Q(y) = a + b\,y$. This is also the market supply function. The market demand function is $x(p) = \frac{c-p}{d}$, where $c > a$ and $b, d \geq 0$. The production of the good causes an externality. Thus, the social marginal costs are $e > 0$ units higher than the marginal costs perceived by the firm, where $c > a + e$

1. Determine the price and quantity in the resulting market equilibrium (without internalization).
   a) In market equilibrium the quantity is $y^M = \frac{a-c}{b+d}$ and the price is $p^M = \frac{ab+cd}{b+d}$.
   b) In market equilibrium the quantity is $y^M = \frac{c-a}{b+d}$ and the price is $p^M = \frac{ad+bc}{b+d}$.
   c) In market equilibrium the quantity is $y^M = \frac{ab+cd}{b+d}$ and the price is $p^M = \frac{a-c}{b+d}$.
   d) In market equilibrium the quantity is $y^M = \frac{a+d}{b+c}$ and the price is $p^M = \frac{ad+bc}{b+c}$.
   e) None of the above answers are correct.

2. Determine the price and quantity in the social optimum (with internalization).
   a) In the social optimum the quantity is $y^{SO} = \frac{c-a-e}{b+d}$ and the price is $p^{SO} = \frac{ad+bc+de}{b+d}$.
   b) In the social optimum the quantity is $y^{SO} = \frac{c-a}{b+d+e}$ and the price is $p^{SO} = \frac{ab+cd+eb}{b+d}$.
   c) In the social optimum the quantity is $y^{SO} = 0$ and the price is undetermined.
   d) In the social optimum the quantity is $y^{SO} = \frac{(a-c)e}{b+d}$ and the price is $p^{SO} = \frac{ab+cd+(b+d)e}{b+d}$.
   e) None of the above answers are correct.
3. Determine the deadweight loss caused by the lack of internalization.
   a) The deadweight loss is $DWL = \frac{e^2}{(b+d)}$.
   b) The deadweight loss is $DWL = \frac{2e}{(b+d)}$.
   c) The deadweight loss is $DWL = \frac{4e^2}{2(b+d)}$.
   d) The deadweight loss is $DWL = \frac{e^2}{2(b+d)}$.
   e) None of the above answers are correct.
4. Determine the social optimum (with internalization) if $c < a + e$.
   a) In the social optimum the quantity is $y^{SO} = \frac{c-a}{b+d}$ and the price is $p^{SO} = \frac{ab+cd+ce}{b+d}$.
   b) In the social optimum the quantity is $y^{SO} = \frac{a-c-e}{b+d+e}$ and the price is $p^{SO} = \frac{ab+cd+de}{b+d+e}$.
   c) In the social optimum the quantity is $y^{SO} = 0$ and the price is indeterminate.
   d) In the social optimum the quantity is $y^{SO} = \frac{(a-c)e^2}{b+d}$ and the price is $p^{SO} = \frac{ab+cd+be}{b+d}$.
   e) None of the above answers are correct.

6.3.1.2 Exercise 2

The production of paper by firm A produces sewage (x, measured in cubic meters) that is dumped into a nearby lake. Profits of the firm in terms of sewage x are given by

$$\Pi_A(x) = 1{,}200\,x - x^2,$$

with $x = 500$ being the maximum possible quantity of sewage. Sewage has a negative influence on the population of fish and thereby harms the profits of firm B, a fishery. Its profits in terms of the absence of sewage y are given by

$$\Pi_B(y) = 500\,y - \frac{1}{2}\,y^2$$

with $y = 500 - x$.

1. Does an externality exist in this situation? If yes, who causes it?
   a) Yes, there is an externality. Firm A causes it.
   b) Yes, there is an externality. Firm B causes it.

   c) The question cannot be answered with the given information.
   d) Yes, there is an externality. The originator cannot be determined, though.
   e) None of the above answers are correct.
2. Determine the marginal profits, $\Pi_A'(x)$ and $\Pi_B'(y)$.
   a) $\Pi_A'(x) = 600 - x$ and $\Pi_B'(y) = y$.
   b) $\Pi_A'(x) = 1{,}200 - 2x$ and $\Pi_B'(y) = 500 - y$.
   c) $\Pi_A'(x) = 600 - x$ and $\Pi_B'(y) = 500 - y$.
   d) $\Pi_A'(x) = 1{,}200 - x$ and $\Pi_B'(y) = 500 - 2y$
   e) None of the above answers are correct.

Firm A has the right to dump sewage into the lake. Firm B can therefore pay firm A to reduce sewage. Assume that the negotiations between the firms cause no transaction costs and that both firms maximize their profits rationally.

3. How much sewage will firm A dump into the lake if it does not negotiate with firm B? What are the total profits of both firms in this situation?
   a) $x = 400$ and $\Pi_A + \Pi_B = 300{,}000$.
   b) $x = 250$ and $\Pi_A + \Pi_B = 25{,}000$.
   c) $x = 500$ and $\Pi_A + \Pi_B = 350{,}000$.
   d) $x = 0$ and $\Pi_A + \Pi_B = 25{,}000$.
   e) None of the above answers are correct.
4. What is the optimal quantity x that maximizes the sum of profits of both firms?
   a) $x = 400$.
   b) $x = 500$.
   c) $x = 0$.
   d) $x = 300$.
   e) None of the above answers are correct.
5. Determine the additional profits compared to a situation without negotiations.
   a) Additional profits amount to 500.
   b) Additional profits amount to 15,000.
   c) Additional profits amount to 25,000.
   d) Additional profits amount to 5,000.
   e) None of the above answers are correct.

6.3.2 Solutions

6.3.2.1 Solutions to Exercise 1

- Question 1, answer b) is correct.
- Question 2, answer a) is correct.
- Question 3, answer d) is correct.
- Question 4, answer c) is correct.

6.3.2.2 Solutions to Exercise 2

- Question 1, answer c) is correct.
- Question 2, answer b) is correct.
- Question 3, answer c) is correct.
- Question 4, answer a) is correct.
- Question 5, answer b) is correct.

7 Decisions and Consumer Behavior

7.1 True or False

7.1.1 Statements

7.1.1.1 Block 1

1. Let $u(x_1, x_2) = x_1 + x_2$ be a utility function. There exists no preference relation which is represented by this utility function.
2. Let $x_1 \succ x_2$ and $x_2 \succ x_3$. Then, the assumption of transitivity implies that $x_1 \succ x_3$.
3. If $u(x_1, x_2) = x_1 \cdot (x_2)^5$ is a utility representation of a preference ordering, then $v(x_1, x_2) = \frac{1}{5} \ln x_1 + \ln x_2$, too, is a utility representation of the same preference ordering.
4. Preferences that fulfill the principle of monotonicity are always convex.

7.1.1.2 Block 2

Assume an individual has income $b > 0$ at his disposal, which he can spend on two goods 1 and 2.

1. A consumer's preference relation is represented by the utility function $u(x_1, x_2) = x_1 \cdot x_2$. Let x_1 be marked on the horizontal axis and x_2 on the vertical axis. If so, the price-consumption path for all $p_1 > 0, p_2 > 0$ is a straight line through the origin with a slope of $\frac{p_1}{p_2}$.
2. For an individual, the two goods are perfect complements. If so, the cross-price elasticity of the Marshallian demand functions always equals zero.
3. The individual's demand for good 1 will decrease if the price of good 1 decreases, provided that x_1 is an inferior good.
4. For the individual, the two goods are perfect substitutes. At the optimum, the demand for one good will always be zero.

M. Kolmar, M. Hoffmann, *Workbook for Principles of Microeconomics*,
Springer Texts in Business and Economics, https://doi.org/10.1007/978-3-319-62662-8_7

7.1.1.3 Block 3

1. A utility function u has a uniquely defined marginal rate of substitution in point x: $MRS_u(x)$. If so, there might exist a utility representation v, which is derived from u through a monotonic transformation and whose marginal rate of substitution at point x, $MRS_v(x)$, differs from the original marginal rate of substitution: $MRS_u(x) \neq MRS_v(x)$.
2. The utility functions of a strictly convex preference relation are characterized by diminishing marginal utility.
3. If the preference ordering of an individual is strictly convex, then its not-worse-than-x sets are convex as well.
4. A preference relation of an individual is called preference ordering if it is complete and transitive.

7.1.1.4 Block 4

Assume a preference relation with utility representation $u(x_1, x_2) = \sqrt{x_1} \cdot x_2$.

1. A transformation according to function $f(u) = u^2$ leads to a utility representation of the same preference relation.
2. A transformation according to function $f(u) = u - 10{,}000$ leads to a utility representation of the same preference relation.
3. The function $v(x_1, x_2) = 0.5 \cdot \ln x_1 + \ln x_2$ is a monotonic transformation of the utility function.
4. The function $v(x_1, x_2) = x_1^2 \cdot x_2^4$ is not a monotonic transformation of the utility function, because the indifference curves are no longer convex due to increasing marginal utility.

7.1.1.5 Block 5

Assume a preference ordering with utility representation $u(x_1, x_2) = c + d \cdot x_1 \cdot x_2$, where $c, d > 0$.

1. The marginal rate of substitution between good 1 and good 2 at point x_1, x_2, $MRS(x_1, x_2)$, is equal to $-\frac{x_2}{x_1}$.
2. The marginal rate of substitution between good 1 and good 2 at point $x_1 = 1, x_2 = 2$, $MRS(x_1, x_2)$, is equal to $\frac{d}{2}$.
3. The underlying preference relation is monotonic.
4. The function $v(x_1, x_2) = \ln x_1 + \ln x_2$ is a monotonic transformation of the above utility function.

7.1.1.6 Block 6

Assume that an individual has a monotonic, strictly convex preference ordering. There are only two goods in the economy. The indifference curves do not touch the axes and have no kinks.

1. If, for a marginal change of the price of good 1, good 1 is a Giffen good, then good 2 has to be a substitute for good 1.

2. If, for a marginal change in income, good 1 is a normal good, then good 2 cannot be a Giffen good.
3. If, for a marginal change of the price, good 1 is a complement to good 2, then good 2 has to be a substitute for good 1.
4. An inferior good can never be ordinary.

7.1.1.7 Block 7

1. An individual with income b has a preference relation represented by the utility function $u(x_1, x_2) = x_1 \cdot x_2$. If so, the cross-price elasticity of the Marshallian demand is always -1.
2. An individual with income b has a preference relation represented by the utility function $u(x_1, x_2) = x_1 + x_2$. If so, the price-consumption path is linear for $p_1 < p_2$.
3. An individual with income b has a preference relation represented by the utility function $u(x_1, x_2) = \min\{x_1, x_2\}$. If so, the price-consumption path for all $p_1 > 0, p_2 > 0$ is a hyperbole.
4. An individual with income b has a preference relation represented by the utility function $u(x_1, x_2) = \sqrt{x_1} + x_2$. In such case, in a solution where the demand for both goods is strictly positive, the demand for x_2 must be constant in income.

7.1.1.8 Block 8

1. Assume an individual with a monotonic preference ordering. The marginal rate of substitution is uniquely defined for every consumption bundle and measures the ratio at which one good can be substituted for another to make the individual indifferent.
2. There exists a utility representation for every preference relation.
3. Assume a utility representation of a preference ordering. Every affine transformation in the form of $v = h \cdot u + f$, where $h > 0$, is a utility representation of the same preference ordering.
4. In a decision theory based on the concept of preference relations, marginal utility measures the increase in utility an individual experiences when he receives an additional (infinitesimally small) unit of a good.

7.1.2 Sample Solutions

7.1.2.1 Sample Solutions for Block 1

1. **False**. This is a case of perfect substitutes. See Figure 7.3 in Chapter 7.1.2.
2. **True**. This is true by definition. See Assumption 7.2 in Chapter 7.1.1.
3. **True**. The absolute values that a utility function assigns to consumption bundles are meaningless. This is the reason why it is called an ordinal concept. Therefore, every monotonic transformation of the utility function $u(x_1, x_2)$ represents

the same preference ordering. See Chapter 7.1.3.

$$v(x_1, x_2) = \frac{\ln u(x_1, x_2)}{5} = \frac{\ln x_1 + 5 \ln x_2}{5} = \frac{1}{5} \ln x_1 + \ln x_2.$$

4. **False**. The assumption of monotonicity implies that the individual prefers larger quantities to smaller quantities. The assumption of convexity implies that individuals prefer balanced to extreme consumption bundles. Here is a counterexample: Let $u(x_1, x_2) = x_1^2 + x_2^2$. Apparently, preferences are non-convex. However, preferences are monotone. See Assumption 7.4 and Assumption 7.5 in Chapter 7.1.1.

7.1.2.2 Sample Solutions for Block 2

1. **True**. This is the description of the income-consumption path. See Figure 7.10 in Chapter 7.2.3. It is derived directly from the optimality condition:

$$MRS(x_1, x_2) = \frac{p_1}{p_2} \Leftrightarrow \frac{x_2}{x_1} = \frac{p_1}{p_2} \Leftrightarrow x_2 = \frac{p_1}{p_2} x_1.$$

The price-consumption-path in this case, however, is a horizontal line (see Figure 7.11a in Chapter 7.2.3).

2. **False**. Let $u(x_1, x_2) = \min\{\alpha x_1, \beta x_2\}$, where $\alpha, \beta > 0$. The Marshallian demand for good 1 is then (see Chapter 7.2.3.3):

$$x_1(\alpha, \beta, b) = \frac{\beta b}{\beta p_1 + \alpha p_2}.$$

It follows that the cross-price elasticity equals (see Definition 14.2 in Chapter 14.3):

$$\epsilon_{p_2}^{x_1} = \frac{dx_1/x_1}{dp_2/p_2} = \frac{\partial x_1}{\partial p_2} \frac{p_2}{x_1} = -\frac{\alpha p_2}{\beta p_1 + \alpha p_2}.$$

3. **False**. Given that x_1 is an inferior good, the demand for it will decrease if the individual's income increases. See Definition 4.4 in Chapter 4.2.
4. **False**. This only holds if $p_1 \neq p_2$.

7.1.2.3 Sample Solutions for Block 3

1. **False**. Utility functions are ordinal concepts and can be defined arbitrarily, as long as preferred alternatives are assigned larger numbers. The marginal rate of substitution does not depend on the arbitrarily chosen utility representation, but is constant for all utility functions representing the same preference ordering. This applies in this case, as the utility function v was derived from u through a monotonic transformation. See Chapter 7.1.3.
2. **False**. The assumption of strict convexity implies only that individuals prefer balanced to extreme alternatives. See Assumption 7.5b in Chapter 7.1.1. Here is a counterexample: Let $u(x_1, x_2) = x_1 + \sqrt{x_2}$. Apparently, preferences are strictly convex. However, the marginal utility $\frac{\partial u}{\partial x_1}$ remains constant.
3. **True**. See Definitions 7.3 and 7.5a in Chapter 7.1.1.
4. **True**. This is true by definition. See Chapter 7.1.1.

7.1.2.4 Sample Solutions for Block 4

Two utility functions represent the same preference ordering if each utility function can be represented as a (positive) monotonic transformation of the other utility function. The function $v(x_1, x_2)$ is a strictly monotonically increasing transformation of u if for arbitrary consumption bundles $\tilde{x} = (\tilde{x}_1, \tilde{x}_2)$ and $\hat{x} = (\hat{x}_1, \hat{x}_2)$ the following holds:

$$u(\tilde{x}) > u(\hat{x}) \Leftrightarrow v(\tilde{x}) > v(\hat{x}).$$

This is the case only if there is a function f, such that:

- f is strictly monotonically increasing at all points for all consumption bundles in the relevant domain, i.e. $f'(X) > 0$ for all $X > 0$;
- $v(x_1, x_2) = f \circ u(x_1, x_2)$, i.e. $v(x_1, x_2) = f(u(x_1, x_2))$.

Please note that x_1 and x_2 cannot be negative. See Chapter 7.2.

1. **True**. Let $f(X) = X^2$, then f is strictly monotonically increasing at all points, since $f'(X) = 2X > 0$ for all $X > 0$. It holds that $v(x_1, x_2) = x_1 \cdot x_2^2 = (u(x_1, x_2))^2 = f(u(x_1, x_2))$. Hence, v is a strictly monotonic transformation of u. Thus, v and u represent the same preference relation.
2. **True**. Let $f(X) = X - 10{,}000$, then f is strictly monotonically increasing at all points, since $f'(X) = 1 > 0$ for all $X > 0$. It holds that $v(x_1, x_2) = \sqrt{x_1} \cdot x_2 - 10{,}000 = u(x_1, x_2) - 10{,}000 = f(u(x_1, x_2))$. Hence, v is a strictly monotonic transformation of u. Thus, v and u represent the same preference relation.
3. **True**. Let $f(X) = \ln(X)$, then f is strictly monotonically increasing at all points, since $f'(X) = \frac{1}{X} > 0$ for all $X > 0$. It holds that $v(x_1, x_2) = 0.5 \cdot \ln(x_1) + \ln(x_2) = \ln(x_1^{0.5}) + \ln(x_2) = \ln(x_1^{0.5} \cdot x_2) = \ln(u(x_1, x_2)) = f(u(x_1, x_2))$. Hence, v is a strictly monotonic transformation of u. Thus, v and u represent the same preference relation.

4. **False**. Let $f(X) = X^4$, then f is strictly monotonically increasing at all points, since $f'(X) = 4\,X^3 > 0$ for all $X > 0$. It holds that $v(x_1, x_2) = x_1^2 \cdot x_2^4 = (u(x_1, x_2))^4 = f(u(x_1, x_2))$. Hence, v is a strictly monotonic transformation of u. Thus, v and u represent the same preference relation.

7.1.2.5 Sample Solutions for Block 5

1. **False**. The marginal rate of substitution between good 1 and good 2 at point x_1, x_2, $MRS(x_1, x_2)$, is defined as:

$$\begin{aligned} MRS(x_1, x_2) &= \left|\frac{dx_2}{dx_1}\right| \\ &= \frac{\partial u/\partial x_1}{\partial u/\partial x_2} = \frac{d \cdot x_2}{d \cdot x_1} = \frac{x_2}{x_1}. \end{aligned}$$

2. **False**. The marginal rate of substitution between good 1 and good 2 at point $x_1 = 1, x_2 = 2$, $MRS(x_1, x_2)$, is defined as:

$$\left|\frac{dx_2}{dx_1}\right| = \frac{x_2}{x_1} = 2.$$

3. **True**. The assumption of monotonicity implies that the individual prefers larger quantities to smaller quantities. The utility function assigns larger numbers to alternatives with a larger quantity x_i, since $\frac{\partial u(x_i, x_j)}{\partial x_i} > 0$. See Assumption 7.4 in Chapter 7.1.1.
4. **True**.

$$\begin{aligned} v(x_1, x_2) &= \ln\,(u(x_1, x_2) - c) - \ln d \\ &= \ln\,(c + d \cdot x_1 \cdot x_2 - c) - \ln d \\ &= \ln d + \ln x_1 + \ln x_2 - \ln d \\ &= \ln x_1 + \ln x_2. \end{aligned}$$

7.1.2.6 Sample Solutions for Block 6

1. **False**. If good 1 is a Giffen good, the demand for it will increase in its price. If good 2 is a substitute for good 1, the demand for it will increase in the price of good 1. However, this is not possible given the budget constraint. Thus, good 2 has to be a complement to good 1. See Definitions 4.2, 4.5, and 4.6 in Chapter 4.2.
2. **False**. If good 2 is a Giffen good, the demand for it will decrease when its price decreases. However, this does not imply anything regarding the effect a change in income would have on the demand for good 2. See Chapter 4.2.
3. **False**. See Figure 7.3 a)–d) and the accompanying explanations in Chapter 7.1.2.
4. **False**. See Chapter 7.2.5 and Chapter 7.2.6.

7.1.2.7 Sample Solutions for Block 7

1. **False**. The optimization problem is as follows:

$$\max_{x_1,x_2} u(x_1,x_2) \text{ s.t. } p_1\,x_1 + p_2\,x_2 = b,$$

with the optimality condition

$$\begin{aligned} MRS(x_1,x_2) &= \frac{p_1}{p_2} \\ \Leftrightarrow \quad \frac{x_2}{x_1} &= \frac{p_1}{p_2} \\ \Leftrightarrow \quad x_2 &= \frac{p_1\,x_1}{p_2}. \end{aligned}$$

We now deal with a system of two equations (optimality condition and budget constraint) and two unknown variables (x_1,x_2), from which we can determine the Marshallian demand function:

$$\begin{aligned} p_1\,x_1 + p_2\,\overbrace{\frac{p_1\,x_1}{p_2}}^{=x_2} &= b \\ \Leftrightarrow x_1 &= \frac{b}{2\,p_1}. \end{aligned}$$

For good 2, it follows that $x_2 = \frac{b}{2\,p_2}$. Hence, the cross-price elasticities are given by:

$$\begin{aligned} \epsilon_{p_2}^{x_1} &= \frac{\partial x_1}{\partial p_2}\frac{p_2}{x_1} = 0 \cdot \frac{2\,p_1\,p_2}{b} = 0 \\ \epsilon_{p_1}^{x_2} &= \frac{\partial x_2}{\partial p_1}\frac{p_1}{x_2} = 0 \cdot \frac{2\,p_1\,p_2}{b} = 0. \end{aligned}$$

See the detailed derivation of the Marshallian demand function in Chapter 7.2.2 as well as Definition 14.2 in Chapter 14.3.

2. **True**. See Figure 7.13 in Chapter 7.2.3.2.
3. **False**. For every combination of income and prices, the individual only demands consumption bundles on the 45° line through the origin. Therefore, the price-consumption path is a straight line. See Chapter 7.2.3.3.
4. **False**. The Marshallian demand function for x_2 is derived analogously to the sample solution to Block 7, Statement 1:

$$\begin{aligned} MRS(x_1,x_2) = \frac{1}{2\,\sqrt{x_1}} &= \frac{p_1}{p_2} \\ \Leftrightarrow x_1 &= \left(\frac{p_2}{2\,p_1}\right)^2. \end{aligned}$$

It follows for the demand of x_2:

$$p_1 \overbrace{\left(\frac{p_2}{2\,p_1}\right)^2}^{=x_1} + p_2\,x_2 = b$$

$$\Leftrightarrow x_2 = \frac{b}{p_2} - \frac{p_2}{4\,p_1}.$$

7.1.2.8 Sample Solutions for Block 8

1. **False**. A counterexample is given by indifference curves with kinks. At the kinks, the MRS is not defined. For an example see Chapter 7.2.3.3.
2. **False**. A counterexample is given by lexicographic preferences. See Chapter 7.1.3.
3. **True**. An affine transformation is a special case of the monotonic transformation.
4. **False**. A preference relation only indicates whether alternative $\tilde{x}$ is better or worse than alternative $\hat{x}$, it does not state how much better or worse. However, that is exactly what is measured by marginal utility. Therefore, the marginal utility has no meaningful interpretation in this setting and is also not invariant to the chosen utility representation.

7.2 Open Questions

7.2.1 Problems

7.2.1.1 Exercise 1

A consumer's preference ordering can be described by the utility function $u(x_1, x_2) = x_1 \cdot x_2$. The goods' prices are given by p_1 and p_2, a consumer's income is represented by b.

1. Determine the Marshallian demand functions for the two goods. Is the consumer free from money illusion?
2. How much will a utility-maximizing consumer demand if $b = 900$, $p_1 = 25$, and $p_2 = 30$.
3. The price of good 1 increases to $p_1 = 36$, while p_2 remains the same. Determine x_1 and x_2 for a utility-maximizing consumer.
4. Comment on the following statement: *A consumer has a Cobb-Douglas utility function, i.e.*

$$u(x_1, x_2) = x_1^{\alpha}\,x_2^{\beta},$$

where $\alpha, \beta > 0$. *In his or her utility maximum, the consumer thus assigns fixed shares of income to each good, which are independent of the goods' prices.*

7.2.1.2 Exercise 2

1. Show that a businessman, who observes a consumer's intransitive preferences, can skim off all of the consumer's income with clever trade offers.
2. A preference relation $\succsim$ is called "weak", as it also includes all indifferent alternatives. A preference relation that only includes the truly preferred alternatives is called "strict", and a preference relation that only includes the indifferent alternatives is called an "indifference relation".
   a) Show how the strict preference relation and the indifference relation can be defined with the help of the weak preference relation.
   b) Do the assumptions of completeness and transitivity also hold for the weak preference relation and the indifference preference relation?
3. Prove for the case of two goods that the marginal rate of substitution for monotonic transformations does not depend on a specific utility representation.
4. Explain why marginal utility cannot be interpreted in a meaningful way within the concept of an ordinal utility theory.

7.2.2 Solutions

7.2.2.1 Solutions to Exercise 1

1. The first-order condition (FOC) is generally given by the condition "marginal rate of substitution = relative price":

$$\begin{aligned} MRS(x_1, x_2) &= \frac{p_1}{p_2} \\ \Leftrightarrow \quad \frac{\partial u/\partial x_1}{\partial u/\partial x_2} &= \frac{p_1}{p_2}. \end{aligned}$$

Then, given our utility function, the following equation holds true:

$$\frac{x_2}{x_1} = \frac{p_1}{p_2}. \tag{7.1}$$

The budget constraint of the individual is

$$p_1 \cdot x_1 + p_2 \cdot x_2 = b.$$

Solving for x_1 leads to

$$x_1 = \frac{b}{p_1} - \frac{p_2}{p_1} \cdot x_2,$$

which can be inserted into the FOC (see Eq. 7.1). This leads to the Marshallian demand function for good 2:

$$\begin{aligned} x_2 &= \frac{p_1}{p_2} \cdot \left(\frac{b}{p_1} - \frac{p_2}{p_1} \cdot x_2 \right) \\ &= \frac{b}{p_2} - x_2 \\ &= \frac{b}{2\,p_2}. \end{aligned}$$

Inserting this Marshallian demand function into the FOC (see Eq. 7.1) leads to the Marshallian demand function for good 1:

$$\begin{aligned} x_1 &= \frac{p_2}{p_1} \cdot \frac{b}{2\,p_2} \\ &= \frac{b}{2\,p_1}. \end{aligned}$$

Absence of Money Illusion. The Marshallian demand functions do not change if all prices and income are increased by the same factor. Examples are inflation (with a corresponding income adjustment) or a currency conversion (see Digression 18 in Chapter 7.2.2).
Let us assume that b, p_1, and p_2 all change by a factor t:

$$\begin{aligned} x_1(t\,p_1, t\,p_2, t\,b) &= \frac{t\,b}{2t\,p_1} = \frac{b}{2\,p_1} = x_1(p_1, p_2, b) \\ x_2(t\,p_1, t\,p_2, t\,b) &= \frac{t\,b}{2t\,p_2} = \frac{b}{2\,p_2} = x_2(p_1, p_2, b). \end{aligned}$$

Therefore, the consumer is free from money illusion.

2. The consumer demands the following quantities:

$$\begin{aligned} x_1(p_1 = 25, p_2 = 30, b = 900) &= \frac{900}{2 \cdot 25} = 18, \\ x_2(p_1 = 25, p_2 = 30, b = 900) &= \frac{900}{2 \cdot 30} = 15. \end{aligned}$$

3. After the price increase, the consumer demands the following quantities:

$$\begin{aligned} x_1(p_1 = 36, p_2 = 30, b = 900) &= \frac{900}{2 \cdot 36} = 12.5, \\ x_2(p_1 = 36, p_2 = 30, b = 900) &= \frac{900}{2 \cdot 30} = 15. \end{aligned}$$

Demand for good 2 does not change as the optimal quantity does not depend on the price of good 1.

4. True. The expenditures for each good only depend on the preferences and income. The consumer devotes $\frac{\alpha}{\alpha+\beta}$ $\left(\frac{\beta}{\alpha+\beta}\right)$ of his or her income to the consumption of good 1 (good 2).

7.2.2.2 Solutions to Exercise 2

1. Assume that a consumer has (intransitive) preferences $x \succ y$, $y \succ z$, and $z \succ x$, with $x, y, z \in X$, but she only possesses alternative x. A vendor now purchases alternatives y and z at arbitrary prices. He then offers z in exchange for x plus a small amount of money, so that the consumer will be better off and accept the offer. Following the same pattern, the vendor then first offers y in exchange for z plus a small amount of money, and after that, x in exchange for y plus a small amount of money. The allocation of alternatives is then identical to the initial situation, in which the consumer possessed alternative x and the vendor possessed alternatives y and z. Yet, the vendor charged a small amount for each trade and a fraction of the consumer's income has thus changed hands. He can therefore repeat this pattern until he has skimmed off the consumer's whole income.
2. a) The statement *alternative x is strictly preferred to alternative y*, with $x, y \in X$, is equivalent to the statement *alternative x is weakly preferred to alternative y and alternative y is not weakly preferred to alternative x*. Formally:

$$x \succ y \quad \Leftrightarrow \quad x \succsim y \quad \wedge \quad \neg(y \succsim x).$$

The statement *alternative x is indifferent to alternative y* is equivalent to the statement *alternative x is weakly preferred to alternative y and alternative y is weakly preferred to alternative x*. Formally:

$$x \sim y \quad \Leftrightarrow \quad x \succsim y \quad \wedge \quad y \succsim x.$$

b) The strict preference relation cannot be complete because each alternative is indifferent to itself. Formally:
A general relation R (e.g. relation $\succsim$, $\succ$, or $\sim$) is complete if $xRy \vee yRx$ holds for all $x, y \in X$. However, $x \succ x$ does not hold.
The indifference relation is not complete, either. Generally, there are alternatives in between in which a strong preference exists.
Transitivity is transferred from the weak preference to both the strict preference and to indifference. We prove this for the strict preference and assume that there are three alternatives with $x \succ y \wedge y \succ z$, with $x, y, z \in X$. Transitivity would imply that $x \succ z$ must hold. From the strict preferences follows the weak preference of the two good bundles, $x \succsim y \wedge y \succsim z$, and thus the weak preference $x \succsim z$ due to transitivity of $\succsim$. It is yet to be shown that $z \succsim x$ cannot hold. If $z \succsim x$ were true, $x \succ y$ would imply $z \succsim y$ due to the transitivity of $\succsim$. However, this contradicts the assumption $y \succ z$, i.e. $z \succsim x \wedge x \succ y \Rightarrow z \succsim y$.

3. The utility derived from the consumption bundle (x_1, x_2) is given by $u = u(x_1, x_2)$. Applying the total differential at point (x_1, x_2) leads to:

$$du(x_1, x_2) = \frac{\partial u(x_1, x_2)}{\partial x_1} dx_1 + \frac{\partial u(x_1, x_2)}{\partial x_2} dx_2 = 0.$$

We solve for $\frac{dx_2}{dx_1}$ to get the slope of the indifference curve:

$$\frac{dx_2}{dx_1} = -\frac{\frac{\partial u(x_1,x_2)}{\partial x_1}}{\frac{\partial u(x_1,x_2)}{\partial x_2}} = -MRS(x_1, x_2).$$

The marginal rate of substitution (MRS) at point (x_1, x_2) is thus given by

$$MRS(x_1, x_2) = \frac{\partial u(x_1, x_2)}{\partial x_1} / \frac{\partial u(x_1, x_2)}{\partial x_2}.$$

Monotonic transformation.
Let $v(x_1, x_2) = f(u(x_1, x_2))$, where $\frac{\partial f(u)}{\partial u} > 0$ for all u.
Applying the total differential leads to:

$$dv(x_1, x_2) = \frac{\partial f(u)}{\partial u} \frac{\partial u(x_1, x_2)}{\partial x_1} dx_1 + \frac{\partial f(u)}{\partial u} \frac{\partial u(x_1, x_2)}{\partial x_2} dx_2 = 0.$$

We solve for $\frac{dx_2}{dx_1}$ to get the slope of the indifference curve:

$$\frac{dx_2}{dx_1} = -\frac{\frac{\partial f(u)}{\partial u} \frac{\partial u(x_1,x_2)}{\partial x_1}}{\frac{\partial f(u)}{\partial u} \frac{\partial u(x_1,x_2)}{\partial x_2}} = -\frac{\frac{\partial u(x_1,x_2)}{\partial x_1}}{\frac{\partial u(x_1,x_2)}{\partial x_2}} = -MRS(x_1, x_2).$$

The MRS of the monotonic transformation $v(x_1, x_2) = f(u(x_1, x_2))$ is thus given by $\frac{\partial u(x_1,x_2)}{\partial x_1} / \frac{\partial u(x_1,x_2)}{\partial x_2}$ and is identical to the MRS of the utility function $u(x_1, x_2)$.
4. A preference relation only reveals whether alternative X is better or worse than alternative Y, but not by how much. Yet, this is exactly what marginal utility would measure. Consequently, it cannot be interpreted in a meaningful way and importantly, it is not invariant to the chosen utility representation.
Let u and v be two utility functions where $v(x_1, x_2) = f(u(x_1, x_2))$, with $\frac{\partial f(u)}{\partial u} > 0$ for all u. Since v is a monotonic transformation of u, the two utility functions represent the same ordinal preferences. The marginal utilities of good 1 are: $\frac{\partial u(x_1,x_2)}{\partial x_1}$ and $\frac{\partial f(u)}{\partial u} \frac{\partial u(x_1,x_2)}{\partial x_1}$. Obviously, these marginal utilities are not generally identical, and thus, the concept of marginal utility cannot be interpreted in a meaningful way within the theory of ordinal utilities.

7.3 Multiple Choice

7.3.1 Problems

7.3.1.1 Exercise 1

Assume a situation with two goods and an income of $b > 0$. How do the following taxes and subsidies affect an individual's budget constraint (refer to Fig. 7.5 in Chapter 7.2.)?

1. A lump-sum tax T which is levied on every person independent of his or her income or market behavior.
   a) The budget constraint rotates outwards at point $(0, \frac{b}{p_2})$.
   b) The budget constraint rotates inwards at point $(0, \frac{b}{p_2})$.
   c) The budget constraint shifts inwards in a parallel way.
   d) The budget constraint shifts outwards in a parallel way.
   e) None of the above answers are correct.
2. A (proportional) quantity subsidy which is levied on the number of consumed units x_1. The total subsidy thus amounts to $s_1 \cdot x_1$, where $s_1 > 0$.
   a) The budget constraint rotates outwards at point $(0, \frac{b}{p_1})$.
   b) The budget constraint rotates outwards at point $(0, \frac{b}{p_2})$.
   c) The budget constraint shifts inwards in a parallel way.
   d) The budget constraint shifts outwards in a parallel way.
   e) None of the above answers are correct.
3. A (proportional) quantity tax which is levied on the number of consumed units x_2. The total tax thus amounts to $t_2 \cdot x_2$, where $t_2 > 0$.
   a) The budget constraint rotates outwards at point $(0, \frac{b}{p_1})$.
   b) The budget constraint rotates inwards at point $(0, \frac{b}{p_1})$.
   c) The budget constraint shifts outwards in a parallel way.
   d) The budget constraint shifts inwards in a parallel way.
   e) None of the above answers are correct.
4. A (proportional) tax which is levied on expenditure $(p_1 \cdot x_1 + p_2 \cdot x_2)$. The total tax thus amounts to $t_A(p_1 \cdot x_1 + p_2 \cdot x_2)$, where $t_A > 0$.
   a) The budget constraint rotates outwards at point $(0, \frac{b}{p_1})$.
   b) The budget constraint rotates inwards at point $(0, \frac{b}{p_1})$.
   c) The budget constraint shifts outwards in a parallel way.
   d) The budget constraint shifts inwards in a parallel way.
   e) None of the above answers are correct.

7.3.1.2 Exercise 2

Determine the marginal rate of substitution $MRS(x_1, x_2)$ at point $(x_1, x_2) = (5, 1)$ for the following utility functions:

1. $u(x_1, x_2) = \ln x_1 + x_2$
   a) 5.
   b) 10.
   c) $\frac{1}{5}$.
   d) -1.
   e) None of the above answers are correct.
2. $u(x_1, x_2) = c \cdot x_1 + d \cdot x_2, c > 0, d > 0$
   a) $c\,d$.
   b) $\frac{c}{d}$.
   c) $\sqrt{c\,d}$.
   d) $c + d$.
   e) None of the above answers are correct.
3. $u(x_1, x_2) = (x_1)^{0.5}(x_2)^{0.5}$
   a) $\frac{5}{4}$.
   b) 5.
   c) $\frac{1}{5}$.
   d) 1.
   e) None of the above answers are correct.
4. $u(x_1, x_2) = \min\{x_1, x_2\}$
   a) 1.
   b) ∞.
   c) 0.
   d) Indeterminate.
   e) None of the above answers are correct.

7.3.1.3 Exercise 3

The prices for good 1 and good 2 are given by p_1 and p_2, respectively, and b represents the budget for which $b \geq p_2$ holds. Determine the Marshallian demand functions for the following utility functions:

1. $u(x_1, x_2) = \min\{\alpha\, x_1, x_2\}, \alpha > 0.$
   a) $x_1 = \frac{b}{p_1 + \alpha\, p_2}, x_2 = \frac{\alpha\, b}{p_1 + \alpha\, p_2}$
   b) $x_1 = x_2 = \frac{b}{\alpha\, p_1 + p_2}$
   c) $x_1 = x_2 = \frac{b}{p_1 + \alpha\, p_2}$
   d) $x_1 = \frac{\alpha}{p_1 + b\, p_2}, x_2 = \frac{\alpha\, b}{p_1 + b\, p_2}$
   e) None of the above answers are correct.
2. $u(x_1, x_2) = \ln x_1 + x_2.$
   a) $x_1 = \frac{p_2}{p_1}, x_2 = \frac{b - p_2}{p_2}$
   b) $x_1 = \frac{b - p_2}{p_2}, x_2 = \frac{p_2}{p_1}$
   c) $x_1 = \frac{p_2}{p_1}, x_2 = \frac{b - 2\, p_2}{2\, p_2}$
   d) $x_1 = x_2 = \frac{b}{p_1 + p_2}$
   e) None of the above answers are correct.

3. $u(x_1, x_2) = \left(\ln\left(x_1{}^{56} x_2{}^{56}\right)\right)^{\frac{1}{7}}$.
   a) $x_1 = \frac{b}{4\,p_1}, x_2 = \frac{b}{4\,p_2}$
   b) $x_1 = \frac{b}{4\,p_2}, x_2 = \frac{b}{4\,p_1}$
   c) $x_1 = x_2 = \frac{b}{p_1 + p_2}$
   d) $x_1 = \frac{b}{2\,p_1}, x_2 = \frac{b}{2\,p_2}$
   e) None of the above answers are correct.

7.3.2 Solutions

7.3.2.1 Solutions to Exercise 1

- Question 1, answer c) is correct.
- Question 2, answer b) is correct.
- Question 3, answer b) is correct.
- Question 4, answer d) is correct.

7.3.2.2 Solutions to Exercise 2

- Question 1, answer c) is correct.
- Question 2, answer b) is correct.
- Question 3, answer c) is correct.
- Question 4, answer d) is correct.

7.3.2.3 Solutions to Exercise 3

- Question 1, answer a) is correct.
- Question 2, answer a) is correct.
- Question 3, answer d) is correct.

8 Costs

8.1 True or False

8.1.1 Statements

8.1.1.1 Block 1

1. The marginal costs intersect the average variable costs at their minimum.
2. The producer surplus is equivalent to profits plus variable costs.
3. In the short run, a firm will supply a positive amount to the market as long as the profits at least cover the fixed costs.
4. The average variable costs of the cost function $C(y) = y^3 + 2y + 10$ are $AVC(y) = y^2 + 2$.

8.1.1.2 Block 2

An entrepreneur produces a good by means of capital and his or her own labor.

1. The economic costs for one unit of his or her own labor equal the maximum wage rate that the entrepreneur could have earned in a different job.
2. The economic costs for one unit of capital can become negative.
3. The economic costs for one unit of capital lower the entrepreneur's profit.
4. The economic costs for one unit of capital are equal to the market interest rate.

8.1.1.3 Block 3

A firm has the cost function $C(y) = y^3 + 50$.

1. The marginal costs are $MC(y) = 2 \cdot 2y^2$.
2. The average costs are $AC(y) = y^2 + \frac{50}{y}$.
3. The average costs are monotonically increasing in y.
4. Average costs and average variable costs are identical for $y \to \infty$.

M. Kolmar, M. Hoffmann, *Workbook for Principles of Microeconomics*,
Springer Texts in Business and Economics, https://doi.org/10.1007/978-3-319-62662-8_8

8.1.1.4 Block 4

1. The function of technological fixed costs is a continuous function.
2. The average cost function cannot be identical to the marginal cost function.
3. The average fixed costs decrease as production increases for all $y > 0$.
4. Fixed costs can be divided into technological and technical fixed costs.

8.1.1.5 Block 5

1. A cost function $C(y_i)$ assigns the minimal costs to every output y_i that are required for i's production.
2. A production function assigns the production-efficient combination of inputs to every output.
3. The marginal product of a production function $y = Y(l)$ measures the change in production y that is caused by an additional unit of an input l.
4. The inverse of the production function defines how much input is needed for producing a specific output if one is producing in a production-efficient manner.

8.1.1.6 Block 6

You are the owner of a firm, which requires only labor to produce apples. The apple tree orchard already exists, and the trees and the land cannot be used for anything else.

1. You employ apple pickers for a monthly salary and with a notice period of a month. Within the time frame of a month, all costs are fixed.
2. You employ apple pickers and pay them according to the amount harvested. All costs are variable.
3. You employ apple pickers on a daily basis for a daily salary. All costs are variable.
4. You harvest apples yourself. The costs are zero.

8.1.2 Solutions

8.1.2.1 Sample Solutions for Block 1

1. **True**. Average variable costs are given by $AVC(y) = VC(y)/y$ (see Definition 8.12 in Chapter 8.2). At the minimum of the average variable costs, $AVC'(y) = 0$ holds. The derivative with respect to y, applying the quotient rule, is:

$$\begin{aligned} AVC'(y) &= 0 \\ \Leftrightarrow \quad \frac{VC'(y) \cdot y - VC(y)}{y^2} &= 0 \\ \Leftrightarrow \frac{VC'(y)}{y} &= \frac{AVC(y)}{y} \\ \Leftrightarrow MC(y) &= AVC(y), \text{ for } y > 0. \end{aligned}$$

2. **False**. The producer surplus is equivalent to profits plus fixed costs. See Chapter 9.3.
3. **False**. In the short run, a firm will produce a positive quantity in the market as long as revenues at least cover the variable costs.
4. **True**. Average variable costs are defined by $AVC(y) = VC(y)/y$ (see Definition 8.12 in Chapter 8.2). Using the given cost function results in

$$AVC(y) = (y^3 + 2y)/y = y^2 + 2.$$

8.1.2.2 Sample Solutions for Block 2

1. **True**. See the concept of opportunity costs in Chapter 1.1 as well as the detailed discussion about opportunity costs in Chapter 8.1.
2. **True**. It corresponds to a negative interest rate.
3. **False**. These costs are zero if the entrepreneur has no alternative occupation. Thus, profits are not necessarily affected.
4. **True**. See Chapter 8.1.

8.1.2.3 Sample Solutions for Block 3

1. **False**.

$$MC(y) = C'(y) = 3y^2.$$

See Definition 8.13 in Chapter 8.2.

2. **True**.

$$AC(y) = \frac{C(y)}{y} = y^2 + \frac{50}{y}.$$

See Definition 8.5 in Chapter 8.2.

3. **False**. The average cost function would be monotonically increasing if the first derivative with regard to y were non-negative for all values of y. This is not true for the function on hand:

$$\frac{d\,AC(y)}{d\,y} = 2y - \frac{50}{y^2} < 0 \Leftrightarrow 2y < \frac{50}{y^2} \Leftrightarrow y < 25^{\frac{1}{3}}.$$

The function is monotonically decreasing in y for $y < 25^{\frac{1}{3}}$.

4. **True**.

$$\lim_{y \to \infty} AC(y) = \lim_{y \to \infty} \left(y^2 + \frac{50}{y}\right) = \infty,$$

$$\lim_{y \to \infty} AVC(y) = \lim_{y \to \infty} y^2 = \infty.$$

8.1.2.4 Sample Solutions for Block 4

1. **False**. See Definition 8.6 in Chapter 8.2.
2. **False**. Assuming that a firm produces with the cost function $C(y) = 2y$, the marginal costs are $MC(y) = C'(y) = 2$ and the average costs are $AC(y) = \frac{C(y)}{y} = 2$. Thus, $MC(y) = AC(y)$. See Chapter 8.2.
3. **True**.

$$AFC(y) = \frac{FC}{y},$$

$$AFC'(y) = -\frac{FC}{y^2} < 0.$$

   See Definitions 8.6, 8.7, and 8.11 in Chapter 8.2.
4. **False**. Fixed costs can either be classified as technological or contractual fixed costs. See Definitions 8.6 and 8.7 in Chapter 8.2.

8.1.2.5 Sample Solutions for Block 5

1. **True**. This is true by definition. See Definition 8.1 in Chapter 8.2.
2. **False**. This would be the inverse of the production function. See Definition 8.2 in Chapter 8.2.
3. **True**. This is true by definition. See Definition 8.3 in Chapter 8.2.
4. **True**. See Chapter 8.2.

8.1.2.6 Sample Solutions for Block 6

1. **True**. This is true by definition. See Definition 8.6 and 8.7 in Chapter 8.2.
2. **True**. This is true by definition. See Definition 8.10 in Chapter 8.2.
3. **False**. They are fixed costs. See Definition 8.6 and 8.7 in Chapter 8.2.
4. **False**. The opportunity costs of the harvest have been disregarded (they can, in extreme cases, be zero, but do not have to be). See Chapter 8.1.

8.2 Open Questions

8.2.1 Problems

8.2.1.1 Exercise 1

1. Assume we have the following production function:

$$Y(l) = l^{\alpha}, \tag{8.1}$$

   with $\alpha > 0$.

a) For which values of α does the production function exhibit a decreasing, constant, or increasing marginal product? Draw the function of the marginal product for $\alpha = \frac{1}{2}$, $\alpha = 1$, and $\alpha = \frac{3}{2}$.
b) Let $w > 0$. Determine the cost function $C(y)$, the marginal cost function $MC(y)$ and the average cost function $AC(y)$ of the above production function.
c) For which values of α do the average cost function and the marginal cost function increase, decrease, or stay constant? Let $w = 1$ and draw both functions for $\alpha = \frac{1}{2}$, $\alpha = 1$, and $\alpha = \frac{3}{2}$.

2. Assume we have the following production function:

$$Y(l) = (l - \gamma)^{\beta}, \tag{8.2}$$

with $\beta \in \left\{\frac{1}{2}, 1\right\}$ and $\gamma > 0$.
a) Let $w > 0$. Determine the cost function $C(y)$, the marginal cost function $MC(y)$ and the average cost function $AC(y)$ of the above production function.
b) For which values of β do the average cost function and the marginal cost function increase, decrease, or stay constant? Let $w = \gamma = 1$ and draw both functions for $\beta = \frac{1}{2}$ and $\beta = 1$.

3. Which of those two production functions is not compatible with a perfectly competitive market (depending on the values of α and β)?

8.2.2 Solutions

8.2.2.1 Solutions to Exercise 1

1. a) Definition 8.3 tells us that the marginal product of a production function measures the change in production y that is caused by an additional unit of an input l. The marginal product is then

$$Y'(l) = \alpha \cdot l^{\alpha-1}.$$

The first derivative of the marginal product function (= second derivative of the production function) tells us, whether the marginal product function is increasing, decreasing or constant in l:

$$Y''(l) = \alpha(\alpha - 1)\, l^{\alpha-2}.$$

Since $\alpha > 0$, the marginal product is

$$\left\{\begin{array}{c}\text{increasing}\\ \text{constant}\\ \text{decreasing}\end{array}\right\} \text{ in } l \text{ if } \alpha \left\{\begin{array}{c}>\\ =\\ <\end{array}\right\} 1.$$

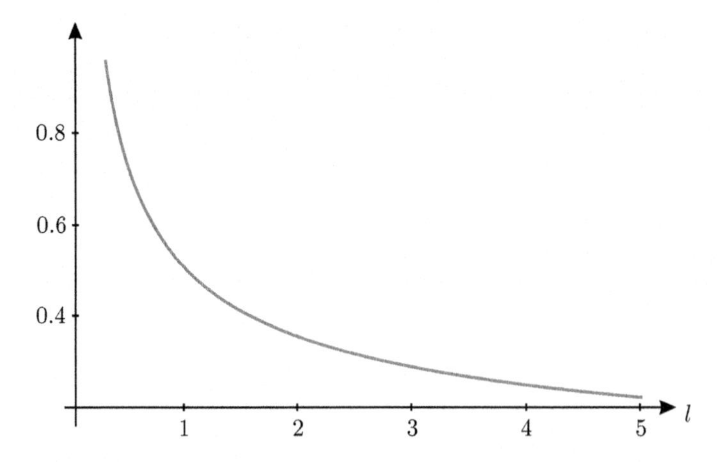

Figure 8.1 Exercise 1.1a) $Y'(l; \alpha = 0.5)$

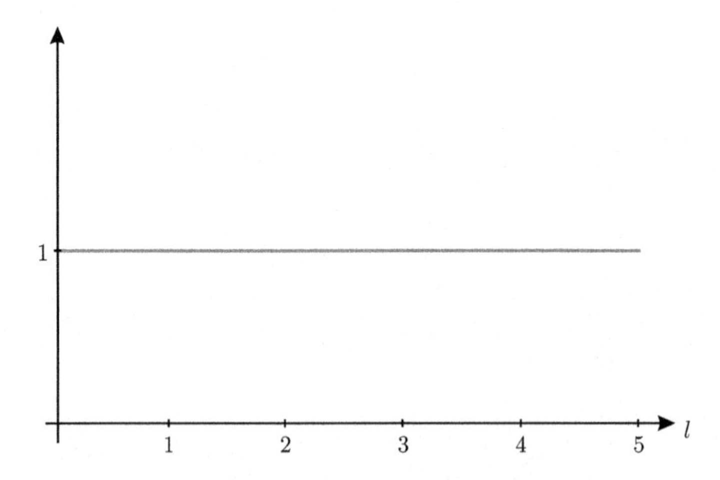

Figure 8.2 Exercise 1.1a) $Y'(l; \alpha = 1)$

Moreover, we need information about the curvature of the marginal product function, i.e. we need the second derivative of the marginal product function (= third derivative of the production function).

$$Y'''(l) = \alpha(\alpha - 1)(\alpha - 2)\, l^{\alpha-3}.$$

Then we get

- $Y'''(l; \alpha = 0.5) > 0$ (convex function),
- $Y'''(l; \alpha = 1) = 0$ (linear function),
- $Y'''(l; \alpha = 1.5) < 0$ (concave function).

These functions are illustrated in Figs. 8.1, 8.2 and 8.3.

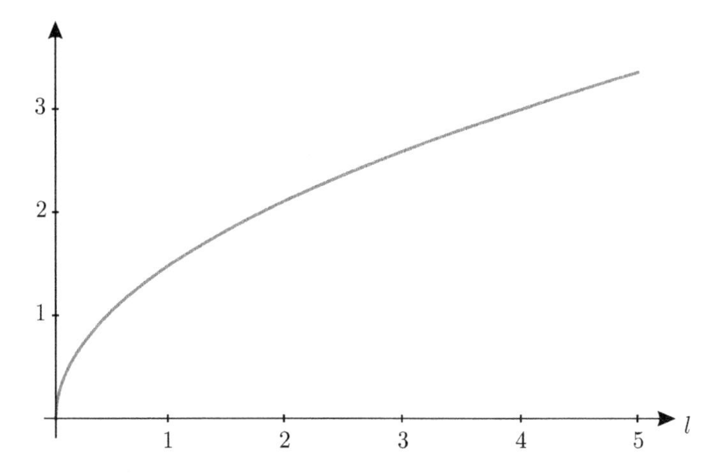

Figure 8.3 Exercise 1.1a) $Y'(l; \alpha = 1.5)$

b) According to Definition 8.4 in Chapter 8.2, the cost function for one-output-one-input technologies $y = Y(l)$ is given by $C(y) = L(y) \cdot w = Y^{-1}(y) \cdot w$. We thus need the inverse of the production function:

$$L(y) = Y^{-1}(y) = y^{\frac{1}{\alpha}}.$$

The cost function is then given by

$$C(y) = w \cdot L(y) = w \cdot y^{\frac{1}{\alpha}}.$$

The marginal cost function then is

$$MC(y) = C'(y) = \frac{w}{\alpha} \cdot y^{\frac{1-\alpha}{\alpha}}, \tag{8.3}$$

and the average cost function is

$$AC(y) = \frac{C(y)}{y} = w \cdot y^{\frac{1-\alpha}{\alpha}}. \tag{8.4}$$

c) The slope of the marginal cost function is

$$MC'(y) = \frac{1-\alpha}{\alpha} \cdot \frac{w}{\alpha} \cdot y^{\frac{1-2\alpha}{\alpha}}.$$

The slope of the average cost function is

$$AC'(y) = \frac{1-\alpha}{\alpha} \cdot w \cdot y^{\frac{1-2\alpha}{\alpha}}.$$

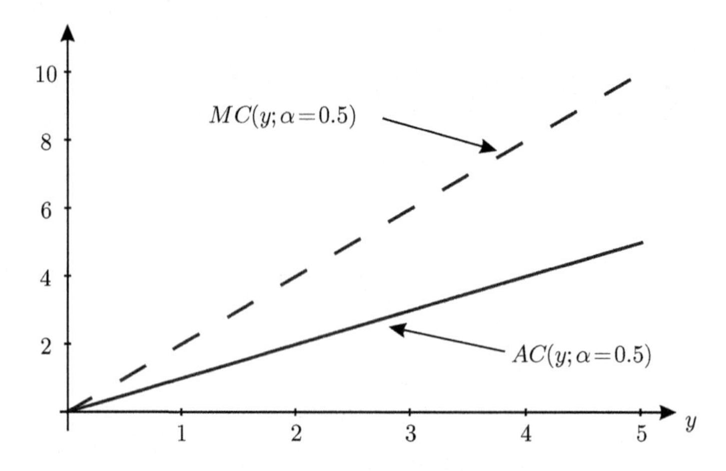

Figure 8.4 Exercise 1.1c) $MC(y;\alpha = 0.5)$ and $AC(y;\alpha = 0.5)$

Since $\alpha > 0$, we get that both marginal costs and average costs are

$$\left\{\begin{array}{c}\text{increasing}\\ \text{constant}\\ \text{decreasing}\end{array}\right\} \text{ if } \alpha \left\{\begin{array}{c}<\\ =\\ >\end{array}\right\} 1.$$

In order to sketch both functions, we need information about (i) the relative position of the functions and (ii) their curvature.

Ad (i) Looking at Eq. 8.3 and Eq. 8.4, we see that $AC(y) = \alpha \cdot MC(y)$ and thus

$$MC(y) \left\{\begin{array}{c}>\\ =\\ <\end{array}\right\} AC(y) \Leftrightarrow \alpha \left\{\begin{array}{c}<\\ =\\ >\end{array}\right\} 1.$$

Ad (ii) The second derivatives of these functions are

$$MC''(y) = \frac{1-2\alpha}{\alpha} \cdot \frac{1-\alpha}{\alpha} \cdot \frac{w}{\alpha} \cdot y^{\frac{1-3\alpha}{\alpha}}.$$

and

$$AC''(y) = \frac{1-2\alpha}{\alpha} \cdot \frac{1-\alpha}{\alpha} \cdot w \cdot y^{\frac{1-3\alpha}{\alpha}}.$$

Consequently, we get

- $AC''(y;\alpha = 0.5) = 0.5 \cdot MC''(y;\alpha = 0.5) = 0$ (linear function),
- $AC''(y;\alpha = 1) = MC''(y;\alpha = 1) = 0$ (linear function),
- $AC''(y;\alpha = 1.5) = 1.5 \cdot MC''(y;\alpha = 1.5) > 0$ (convex function).

These functions are illustrated in Figs. 8.4, 8.5 and 8.6.

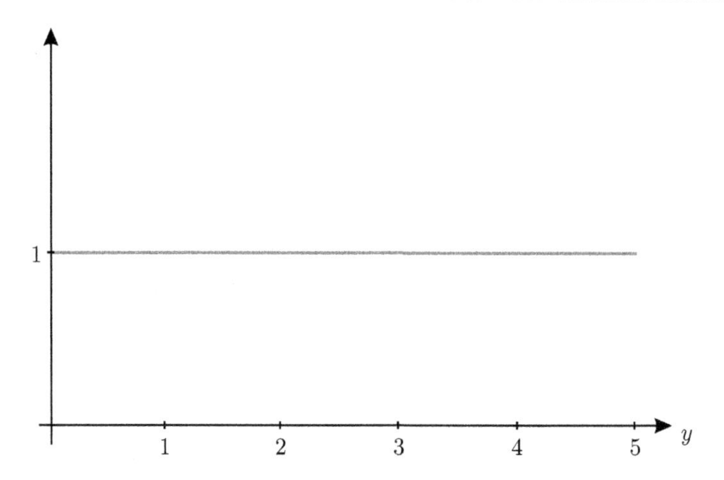

Figure 8.5 Exercise 1.1c) $MC(y; \alpha = 1)$ and $AC(y; \alpha = 1)$

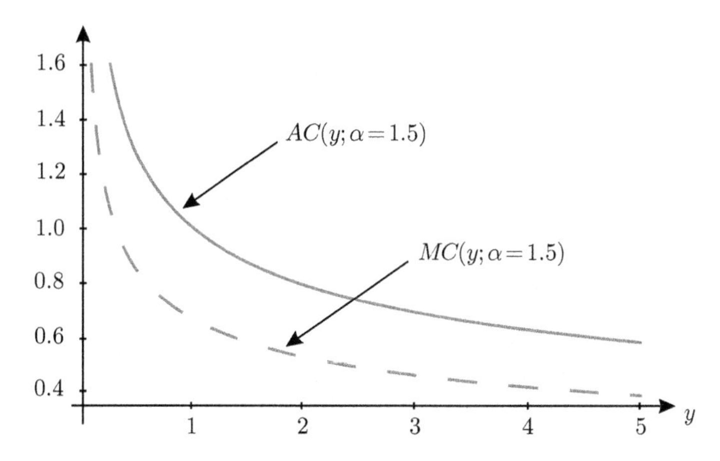

Figure 8.6 Exercise 1.1c) $MC(y; \alpha = 1.5)$ and $AC(y; \alpha = 1.5)$

2. a) Again, we need the inverse of the production function:

$$L(y) = Y^{-1}(y) = \gamma + y^{\frac{1}{\beta}}.$$

The cost function then becomes:

$$C(y) = w \cdot L(y) = \underbrace{w \cdot \gamma}_{=FC} + \underbrace{w \cdot y^{\frac{1}{\beta}}}_{=VC(y)}.$$

The marginal cost function $MC(y)$ is

$$MC(y) = C'(y) = \frac{w}{\beta} \cdot y^{\frac{1-\beta}{\beta}}, \tag{8.5}$$

and the average cost function $AC(y)$

$$AC(y) = \frac{C(y)}{y} = \underbrace{\frac{w \cdot \gamma}{y}}_{=AFC(y)} + \underbrace{w \cdot y^{\frac{1-\beta}{\beta}}}_{=AVC(y)}. \tag{8.6}$$

b) The slope of the marginal cost function is

$$MC'(y) = \frac{1-\beta}{\beta} \cdot \frac{w}{\beta} \cdot y^{\frac{1-2\beta}{\beta}}.$$

The slope of the average cost function is

$$AC'(y) = -\frac{w \cdot \gamma}{y^2} + \frac{1-\beta}{\beta} \cdot w \cdot y^{\frac{1-2\beta}{\beta}}.$$

Thus, we get

- $MC'(y; \beta = 0.5) = 2 \cdot w \cdot y^0 = 2 \cdot w > 0$,
- $AC'(y; \beta = 0.5) = -\frac{w \cdot \gamma}{y^2} + w \cdot y^0 = -\frac{w \cdot \gamma}{y^2} + w$,
- $MC'(y; \beta = 1) = 0$,
- $AC'(y; \beta = 1) = -\frac{w \cdot \gamma}{y^2} < 0$.

Consequently, $MC(y; \beta = 0.5)$ is a monotonically increasing, $AC(y; \beta = 1)$ a monotonically decreasing, and $MC(y; \beta = 1)$ a constant function. $AC(y; \beta = 0.5)$, however, is a non-monotonic function.
In order to sketch these functions, we need information about (i) their curvature and (ii) the relative position of these functions.
Ad (i) The second derivatives are

- $MC''(y; \beta = 0.5) = 0$ (linear function),
- $AC''(y; \beta = 0.5) = \frac{2 \cdot w \cdot \gamma}{y^3} > 0$ (convex function),
- $MC''(y; \beta = 1) = 0$ (linear function),
- $AC''(y; \beta = 1) = \frac{2 \cdot w \cdot \gamma}{y^3} > 0$ (convex function).

Ad (ii) Looking at Eq. 8.5 and Eq. 8.6 shows us that

$$MC(y; \beta = 0.5) \begin{Bmatrix} < \\ = \\ > \end{Bmatrix} AC(y; \beta = 0.5)$$

$$\Leftrightarrow 2 \cdot w \cdot y \begin{Bmatrix} < \\ = \\ > \end{Bmatrix} w \cdot y + \frac{w \cdot \gamma}{y}$$

$$\Leftrightarrow y \begin{Bmatrix} < \\ = \\ > \end{Bmatrix} \sqrt{\gamma}.$$

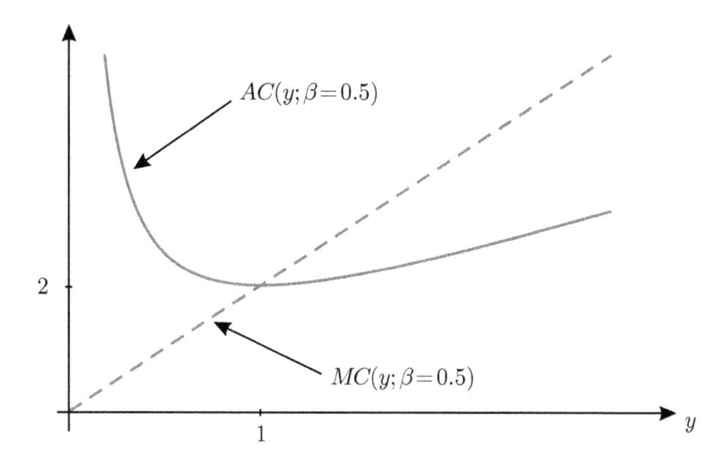

Figure 8.7 Exercise 1.2b) $MC(y; \beta = 0.5)$ and $AC(y; \beta = 0.5)$

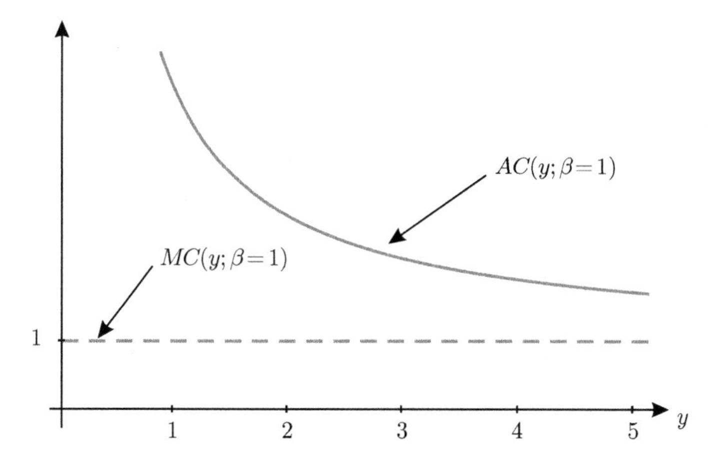

Figure 8.8 Exercise 1.2b) $MC(y; \beta = 1)$ and $AC(y; \beta = 1)$

Since $AVC(y; \beta = 1) = MC(y; \beta = 1) = w$ and $AFC(y; \beta = 1) = \frac{w \cdot \gamma}{y} > 0$, we have $AC(y; \beta = 1) > MC(y; \beta = 1)$. These functions are illustrated in Figs. 8.7 and 8.8.

3. Monotonically decreasing average costs are incompatible with perfect competition (see Chapter 9.2). Thus, the production function (8.1) at $\alpha > 1$ and the production function (8.2) at $\beta = 1$ are incompatible with perfect competition. In the first case, decreasing average costs result from a productivity increase in the factor of production "labor": the more labor one uses, the larger the marginal productivity of labor will be, i.e., the larger the additional output will be. In the second case, decreasing average costs result from the combination of constant marginal costs and positive fixed costs: marginal costs always correspond to av-

erage variable costs, which in turn correspond to average wage costs (w). This is true regardless of the output. However, the larger the output, the lower the average fixed costs ($AFC(y)$), because the fixed costs can be distributed among more units of the output. Thus, total average costs ($AC(y)$) are monotonically decreasing for $y > 0$.

8.3 Multiple Choice

8.3.1 Problems

8.3.1.1 Exercise 1

A firm has fixed short-run capital costs of 16 units of money. It produces y units of a good using capital and labor. With its existing capital stock, it can produce $y = l^a$ units of the good ($a > 0$) with l units of labor. The wage rate per unit of labor is $w = 2$.

1. Calculate the firm's short-term cost function.
   a) $C(y) = 2y^a + 16$.
   b) $C(y) = 2y^a$
   c) $C(y) = 16$.
   d) $C(y) = 2y^{\frac{1}{a}} + 16$.
   e) None of the above answers are correct.
2. Calculate the firm's variable, fixed, average, and marginal costs.
   a) $VC(y) = 2y^a$ and $FC = 16$ and $AC(y) = 2y^{a-1} + \frac{16}{y}$ and $MC(y) = 2ay^{a-1}$.
   b) $VC(y) = 2y^{\frac{1}{a}}$ and $FC = 16$ and $AC(y) = 2y^{\frac{1-a}{a}} + \frac{16}{y}$ and $MC(y) = \frac{2}{a}y^{\frac{1-a}{a}}$.
   c) $VC(y) = 2y^a$ and $FC = 0$ and $AC(y) = 2y^{a-1}$ and $MC(y) = 2ay^{a-1}$.
   d) $VC(y) = 2y$ and $FC = 16$ and $AC(y) = 2y + \frac{16}{y}$ and $MC(y) = 2$.
   e) None of the above answers are correct.
3. Now, assume that $a < 1$. The good's price is p. Determine the quantity supplied, for which "price = marginal costs" holds.
   a) $y(p) = \left(\frac{p}{2a}\right)^{\frac{1}{a-1}}$.
   b) $y(p) = \left(\frac{a\,p}{2}\right)^{\frac{a}{1-a}}$.
   c) $y(p) = 2\,p^{\frac{1}{a}}$.
   d) The quantity offered is undetermined since marginal costs are constant.
   e) None of the above answers are correct.

8.3.1.2 Exercise 2

A firm's production function is $Y(l) = 10\sqrt{l}$, where y is the quantity of goods and l is labor. The firm additionally has technological fixed costs of $TFC = 50$. The good's market price is $p = 5$.

1. How much labor l has the firm used in order to produce 20 units of the good?
   a) $L(y = 20) = 4$.
   b) $L(y = 20) = 12$.
   c) $L(y = 20) = 5.5$.
   d) $L(y = 20) = 8$.
   e) None of the above answers are correct.
2. The wage is $w = 2$. Determine the firm's cost function for $y > 0$.
   a) $C(y) = 50 + 3y$.
   b) $C(y) = 50$.
   c) $C(y) = \frac{y^2}{50} + 50$.
   d) $C(y) = 4y^2 + 50$.
   e) None of the above answers are correct.

8.3.1.3 Exercise 3

In an industry, all firms are producing a homogeneous good. With l units of labor they can produce $Y(l) = 100\sqrt{l}$ units of the good. The wage is $w = 100$, and the firms have contractual fixed costs of 100.

1. Determine the firms' cost functions.
   a) The cost functions are $C(y) = \frac{y^2}{100} + 100$.
   b) The cost functions are $C(y) = 100y^2 + 100$.
   c) The cost functions are $C(y) = \frac{\sqrt{y}}{100} + 100$.
   d) The cost functions are $C(y) = 50y + 100$.
   e) None of the above answers are correct.
2. Determine the average and marginal-cost functions.
   a) $AC(y) = 100y + \frac{100}{y}$ and $MC(y) = \frac{2y}{100} + \frac{100}{y}$.
   b) $AC(y) = \frac{y}{100} + \frac{100}{y}$ and $MC(y) = \frac{2y}{100} + \frac{100}{y}$.
   c) $AC(y) = 50 + \frac{100}{y}$ and $MC(y) = 50$.
   d) $AC(y) = \frac{y}{100} + \frac{100}{y}$ and $MC(y) = \frac{2y}{100}$.
   e) None of the above answers are correct.

8.3.2 Solutions

8.3.2.1 Solutions to Exercise 1

- Question 1, answer d) is correct.
- Question 2, answer b) is correct.
- Question 3, answer b) is correct.

8.3.2.2 Solutions to Exercise 2

- Question 1, answer a) is correct.
- Question 2, answer c) is correct.

8.3.2.3 Solutions to Exercise 3

- Question 1, answer a) is correct.
- Question 2, answer d) is correct.

A Second Look at Firm Behavior Under Perfect Competition 9

9.1 True or False

9.1.1 Statements

9.1.1.1 Block 1

Assume a profit-maximizing firm.

1. Assume that the firm supplies a strictly positive and finite quantity. Then, the rule "marginal revenues = marginal costs" holds in the optimum.
2. A firm in perfect competition always supplies according the rule "price = marginal costs" if the resulting revenues at least cover the average variable costs.
3. The firm will never make losses in its optimum because it can avoid these by leaving the market.
4. In the long-run market equilibrium with free market entry and exit, a firm's producer surplus is always equal to zero.

9.1.1.2 Block 2

Assume a profit-maximizing firm with a cost function of $C(y) = y^2 + 49$ in a market with perfect competition.

1. The average costs of this firm are equal to the marginal costs at the minimum of the average cost curve.
2. The average variable costs are $AVC(y) = 2y + \frac{49}{y}$.
3. Assume that the firm only produces with one factor (labor), l. The wages are $w = 4$. That means that the production function of the firm is $y = 4 \cdot l^{\frac{1}{2}}$.
4. Assume that all firms are symmetric. In the long-run market equilibrium with perfect competition and with free market entry and exit, the equilibrium price is $p = 14$.

M. Kolmar, M. Hoffmann, *Workbook for Principles of Microeconomics*,
Springer Texts in Business and Economics, https://doi.org/10.1007/978-3-319-62662-8_9

9.1.1.3 Block 3

A firm has the cost function $C(y) = y^a$.

1. Let $a > 1$. In perfect competition, the firm's supply function, at a market price of p, is equal to $y(p) = \left(\frac{p}{a}\right)^{\left(\frac{1}{a-1}\right)}$.
2. If $a > 1$, then a market with perfect competition cannot function if all firms are symmetric.
3. If $a = 1$, then in perfect competition, the firm's optimal supply is any (weakly) positive quantity.
4. If $a < 1$, then the firm is producing with decreasing average costs.

9.1.1.4 Block 4

1. Profits correspond to the producer surplus if fixed costs are zero.
2. Assume technological fixed costs. In the long run, a firm will supply a positive quantity as long as the price is larger than or equal to the average costs.
3. The minimum of the marginal costs always corresponds to the average costs at that point.
4. If an industry produces with constant marginal costs and fixed costs equal to zero, then the number of firms supplying a positive quantity in the long-run market equilibrium with perfect competition and free market entry and exit is uniquely defined.

9.1.1.5 Block 5

1. If a firm supplies a positive quantity to a market with perfect competition, then the price must be greater or equal to the average costs.
2. If a firm produces with constant marginal costs and positive fixed costs, its average costs are always higher than its marginal costs.
3. The price corresponds to a firm's marginal revenues in a market equilibrium under perfect competition.
4. The cost function of a firm in a market with perfect competition is $C(y) = 0.5 \cdot y^2$. Therefore, the supply function is $y(p) = 2p$.

9.1.1.6 Block 6

Assume a strictly convex cost function.

1. In the long run, if a firm is making losses, then it should always leave the market.
2. In the short run, the supply function of a firm always corresponds to the inverse of the marginal costs function.
3. In the long-run market equilibrium with free market entry and exit, all firms supply at the minimum of their marginal costs.
4. If the market price is lower than the average variable costs, the firm should withdraw from the market even in the short run.

9.1.1.7 Block 7

1. Contractual fixed costs are irrelevant for the firm's supply during the term contract.
2. If the average costs are a decreasing function of the quantity produced, then a market with perfect competition and free market entry and exit cannot function.
3. The technological fixed costs are zero if a firm is not producing anything.
4. If there is scarcity, then the marginal costs of production cannot decrease.

9.1.1.8 Block 8

Assume that there are 10,000 licenses for taxis in a market, which are all in use. The costs for running a taxi creates an identical u-shaped average-cost function in the number of trips. The market demand for taxi services is decreasing, and the taxi market is in a long-run equilibrium with the 10,000 licenses and with perfect competition and free market entry and exit.

1. Due to immigration, the demand for taxi services increases. Thus, *ceteris paribus*, the equilibrium price increases both in the short and in the long run.
2. *Uber* enters the market, enabling private individuals to offer taxi services. *Ceteris paribus*, this has no long-run effect on the equilibrium price if the private *Uber* drivers have the same average-cost function as the taxi drivers.
3. *Uber* enters the market, enabling private individuals to offer taxi services. *Ceteris paribus*, this will force all taxi drivers to leave the market if the private individuals have lower average cost for any number of trips (= output) than the taxi drivers do. (Assume that the supply of *Uber* drivers is sufficiently large.)
4. The city introduces a road toll within the city for private passenger cars. *Ceteris paribus*, the equilibrium price for taxi services will increase in both the long run and the short run.

9.1.2 Solutions

9.1.2.1 Sample Solutions for Block 1

1. **True**. The firm acts as a price taker and chooses its profit-maximizing quantity:

$$\max_y \pi(y) = \max_y p \cdot y - C(y).$$

The first order condition then leads to

$$\pi'(y) = \underbrace{p}_{MR(y)} - \underbrace{C'(y)}_{MC(y)} = 0$$
$$\Leftrightarrow MR(y) = MC(y).$$

See the detailed discussion in Chapter 9.1.

2. **False**. At the very least, the variable costs $VC(y)$ have to be covered by the revenues. This implies that the price (not the revenues) has to at least cover the average variable costs:

$$\begin{aligned} p\,y - VC(y) &\geq 0 \\ \Leftrightarrow p &\geq \frac{VC(y)}{y} = AVC(y). \end{aligned}$$

   See Chapter 9.3.
3. **False**. In the short run, it may be rational to stay in business even if the firm incurs losses. In particular, this can be the case if there are contractual fixed costs. In this case, it is still possible to earn a positive producer surplus if at least the variable costs are covered. See Chapter 9.3.
4. **False**. In the long run market equilibrium with free market entry and exit, a firm's profit is always equal to zero.

9.1.2.2 Sample Solutions for Block 2

1. **True**. The following holds at the minimum of the average costs for $y > 0$ (taking the quotient rule into account):

$$\begin{aligned} AC'(y) &= 0 \\ \Leftrightarrow \quad \frac{C'(y) \cdot y - C(y)}{y^2} &= 0 \\ \Leftrightarrow \frac{C'(y)}{y} &= \frac{AC(y)}{y} \\ \Leftrightarrow MC(y) &= AC(y). \end{aligned}$$

   See Chapter 9.3.
2. **False**. The average variable costs are $AVC(y) = \frac{VC(y)}{y} = \frac{y^2}{y} = y$. See Definition 8.12 in Chapter 8.2.
3. **False**. If the firm's production function were $Y(l) = 4 \cdot \sqrt{l}$, then $L(y) = \frac{y^2}{16}$ and $C = w \cdot L(y) = w \cdot \frac{y^2}{16}$. Thus, $C(y) = \frac{y^2}{4}$. The correct production function can be derived from the following observation:

$$\begin{aligned} C(y) = w \cdot L(y) &= 49 + y^2 \\ \Leftrightarrow \quad 4 \cdot l - 49 &= y^2 \\ \Leftrightarrow \quad Y(l) &= \sqrt{4l - 49}. \end{aligned}$$

   Thus, the costs at the point $l = 12.25$ correspond to the value $4 \cdot 12.25 = 49$ (the fixed costs), and the output is zero.
4. **True**. All firms produce in the minimum of the average costs function (zero profit) in the long-run equilibrium, with free market entry and exit. See Chapter

9.3. Thus, remembering that $AC(y) = MC(y)$ at the minimum of the average-cost function, it follows that:

$$\begin{aligned}\frac{C(y)}{y} &= C'(y)\\ \Leftrightarrow y + \frac{49}{y} &= 2y\\ \Leftrightarrow 49 &= y^2\\ \Leftrightarrow 7 &= y^*.\end{aligned}$$

The value of the average costs or marginal costs corresponds to the market price at y^*, i.e. $P(y^*) = AC(y^*) = MC(y^*) = 14$.

9.1.2.3 Sample Solutions for Block 3

1. **True**. "Marginal revenues = marginal costs" is applicable, in which marginal revenues correspond to p in perfect competition. Thus, in optimum:

$$p = MC(y) \Leftrightarrow p = a \cdot y^{a-1} \Leftrightarrow y^{a-1} = \frac{p}{a} \Leftrightarrow y(p) = \left(\frac{p}{a}\right)^{\frac{1}{a-1}}.$$

2. **False**. If $a > 1$, then the firm is producing with monotonically increasing marginal costs:

$$\begin{aligned}MC(y) &= C'(y) = a \cdot y^{a-1}\\ \Rightarrow MC'(y) &= \underbrace{(a-1) \cdot a}_{>0}\, y^{a-2} > 0.\end{aligned}$$

Increasing marginal costs are not an exclusion criterion for perfect competition, but characteristic for this type of market.

3. **True**. If $a = 1$, the average costs are $AC(y) = \frac{C(y)}{y} = \frac{y}{y} = 1$. If the price is $p = 1$, then the average costs are exactly covered and the firm can produce as much of the good as it wishes to. See the discussion about constant marginal costs in Chapter 9.2.
4. **True**.

$$\begin{aligned}AC(y) &= \frac{C(y)}{y} = y^{a-1}\\ \Rightarrow AC'(y) &= \underbrace{(a-1)}_{<0}\, y^{a-2} < 0.\end{aligned}$$

The average costs decrease as the quantity produced increases.

9.1.2.4 Sample Solutions for Block 4

1. **True**. The producer surplus is the aggregate difference between the market price and the supplier's minimum willingness to sell. This corresponds to revenues minus variable costs, which furthermore corresponds to profits if fixed costs are zero. See Definition 5.5 in Chapter 5.2 and Chapter 9.3.
2. **True**. In the long run, if all costs can be avoided by not entering the market, the optimal quantity supplied is determined by the rule "price = marginal costs" if the market price is greater than (or equal to) the average costs. If this is not the case, the firm will not be active in the market. See Chapter 9.3.
3. **False**. The marginal-cost function intersects the average-cost function in its minimum, i.e., the minimum of the average-cost function corresponds to the marginal costs at this point, and not vice versa. See Chapter 9.3.
4. **False**. A technology with constant marginal costs represents a special case in the functioning of competitive markets. Constant marginal costs imply that the price is either strictly larger, strictly smaller or equal to the marginal costs, regardless of the quantity supplied. Therefore, the optimal quantity supplied is either infinite, zero, or undefined. See Chapter 9.2.

9.1.2.5 Sample Solutions for Block 5

1. **False**. The statement is only true in the long run. In the short run, a firm will supply a positive amount if the price is equal or greater than the minimum of the average variable costs. See Chapter 9.3.
2. **True**. Suppose $C(y) = a\,y + FC$, where $a > 0$ and $FC > 0$. Then, marginal costs are given by $MC(y) = C'(y) = a$, and average costs by $AC(y) = \frac{C(y)}{y} = \frac{a\,y+FC}{y} = a + \frac{FC}{y}$. Since $FC > 0$, $AC = a + \frac{FC}{y} > MC(y) = a$ holds.
3. **True**. In a market with perfect competition, all firms act as price takers. Revenues of such a firm are thus given by $R(y) = p \cdot y$. Marginal revenues, defined as the derivative of the revenue function with respect to y, then amount to $MR(y) = R'(y) = p$. See Definition 9.1 in Chapter 9.1.
4. **False**. Under perfect competition, the individual supply function corresponds to the inverse of the marginal-cost function if the price is larger than the average costs (and zero else). $MC(y) = y$ and $AC(y) = 0.5y$. Because the price corresponds to the marginal costs, $p = y > AC(y)$ holds. Thus, the supply function is $y(p) = p$, for $p > 0$. See Chapter 9.3.

9.1.2.6 Sample Solutions for Block 6

1. **False**. A firm that faces contractual fixed costs will supply a positive quantity as long as the price is larger than or equal to the average variable costs. See Chapter 9.3.
2. **False**. Depending on the time horizon (short or long run) and the type of fixed costs (TFC or CFC), average costs or average variable costs have an impact on the supply function. See Chapter 9.3.

3. **False**. All firms sell at the minimum of their average costs. If firms are making zero profits, they are producing the good at minimal average costs because the marginal- and average-cost functions intersect at the average costs' minimum. See Chapter 9.4.
4. **True**. See sample solution to Block 6, Statement 1.

9.1.2.7 Sample Solutions for Block 7

1. **True**. The contractual fixed costs of production are the costs that are independent of production. See Definition 8.7 in Chapter 8.2.
2. **True**. Competitive markets cannot function if average costs are monotonically decreasing. In industries with such technologies, firms with larger market shares have an advantage, because they can produce at lower average costs and, hence, can sell at lower prices compared to their smaller competitors. Therefore, a large market share protects firms to some extent against competition. See Chapter 9.2.
3. **True**. The technological fixed costs are the costs that occur once a firm starts production, but which then are independent of the volume of production. See Definition 8.6 in Chapter 8.2.
4. **False**. A counter example is a natural monopoly. A special case of such a technology is discussed in Chapter 6, where it is argued that club goods are sometimes called natural monopolies, because they imply decreasing average costs by increasing the number of users. See also sample solution to Block 7, Statement 2.

9.1.2.8 Sample Solutions for Block 8

1. **True**. Because the marginal costs increase to the right of the minimum, and the number of taxis does not increase, the price and the quantity will increase in the new equilibrium. See Chapter 9.3.
2. **True**. The equilibrium price remains unchanged because it remains in the minimum of the average costs. See Chapter 9.3.
3. **True**. Due to their different cost structure, the *Uber* drivers can offer their service at a lower price and will thus supply the entire market. See Chapter 9.3.
4. **True**. Due to the road toll, the taxi drivers' fixed costs will increase, which means that the minimum of the average costs and, thus, the market price will increase. See Chapter 9.3.

9.2 Open Questions

9.2.1 Problems

9.2.1.1 Exercise 1

Figure 9.1 represents the functions of the marginal costs, $MC(y)$, average costs, $AC(y)$, and the average variable costs, $AVC(y)$, of a firm operating in a polypoly.

1. Let the market price be $\bar{p} = 10$. In the figure, illustrate two different ways to determine the producer surplus in the short-run optimum.
2. Sketch the average-fixed-costs function, $AFC(y)$.
3. Assume that all firms participating in this market have identical cost functions. What is the market price in the long-run market equilibrium with free entry and exit? What is the supply of a single firm? Illustrate the total costs at the equilibrium in a graph.
4. Determine the cost function $C(y)$. Assume that all functions in Fig. 9.1 are continuous for $y > 0$.
5. Determine the long-run and short-run supply function of the firm and differentiate between technological and contractual fixed costs.
6. Labor (l) is the only input factor, and its price is $w = 2$. What is the production function of the firm, $Y(l)$? Sketch the production function.

9.2.1.2 Exercise 2

Assume there are 1,000 booths selling Bavarian veal sausages in the long-run market equilibrium with free entry and exit. Every booth has linearly increasing marginal costs and strictly positive technological fixed costs. The market demand function for veal sausages has a normal shape.

1. Describe the equilibrium for both, an individual firm and the total market with the help of two graphs.
2. The city commands that there are only be 800 licences for veal-sausage-booths. How will this influence the market and the vendors in the market? Illustrate this with the help of two graphs.

9.2.2 Solutions

9.2.2.1 Solutions to Exercise 1

1. In the optimum y^*, $p = MC(y^*)$ holds for each firm. In the short run, producer surplus $PS(y^*)$ must be weakly positive, i.e.

$$PS(y^*) \geq 0 \Leftrightarrow p \cdot y^* - VC(y^*) \geq 0 \Leftrightarrow p \geq AVC(y^*).$$

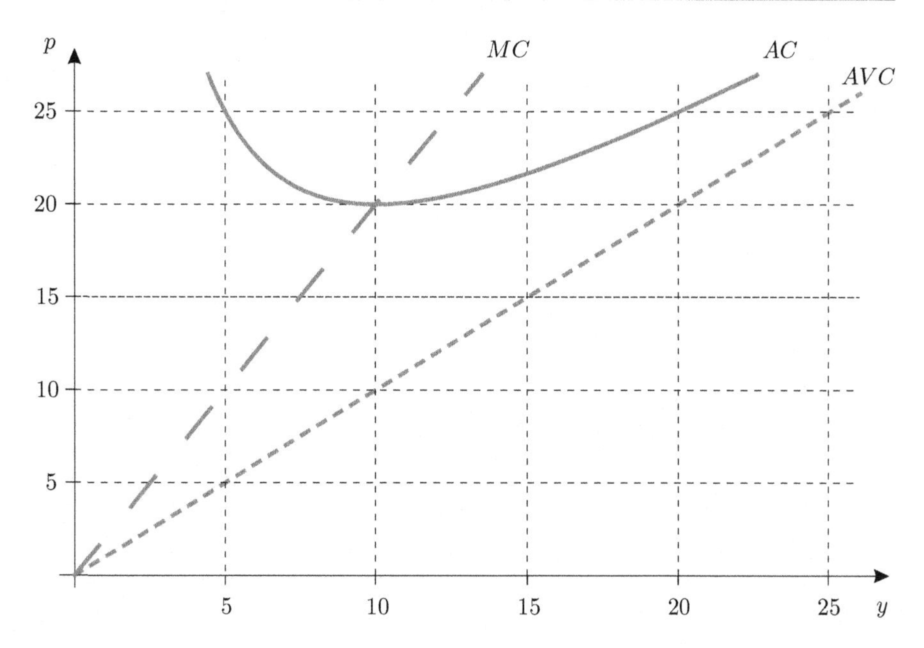

Figure 9.1 Exercise 1

Obviously, this condition is fulfilled for $\bar{p} = 10$ and $y^* = 5$ (see intersection of price and marginal-costs curve in Fig. 9.2).

In order to illustrate the producer surplus at y^*, the difference between revenues $(\bar{p} \cdot y^*)$ and variable costs $(VC(y^*))$ needs to be shown. There are two ways ways to determine the variable costs:

a) Variable costs at y^* correspond to average variable costs at the point y^* multiplied by the quantity (y^*):

$$VC(y^*) = AVC(y^*) \cdot y^*.$$

b) Variable costs at y^* are equivalent to the area underneath the marginal-cost function up to y^*:

$$VC(y^*) = \int_{y=0}^{y=y^*} MC(y)\,\mathrm{d}y.$$

Thus, there are also two ways to illustrate the difference between revenues and variable costs (see Figs. 9.2 and 9.3).

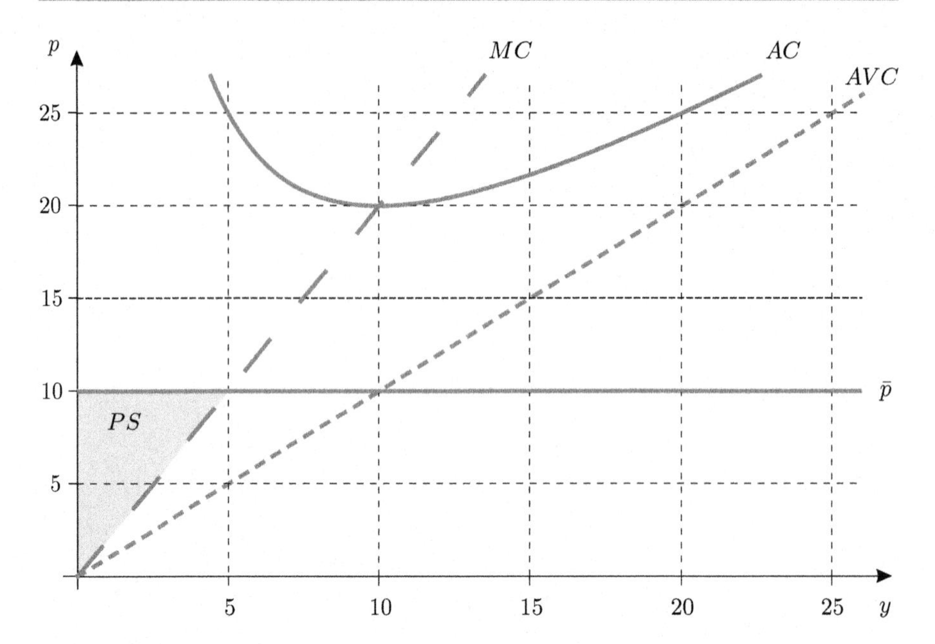

Figure 9.2 Exercise 1.1

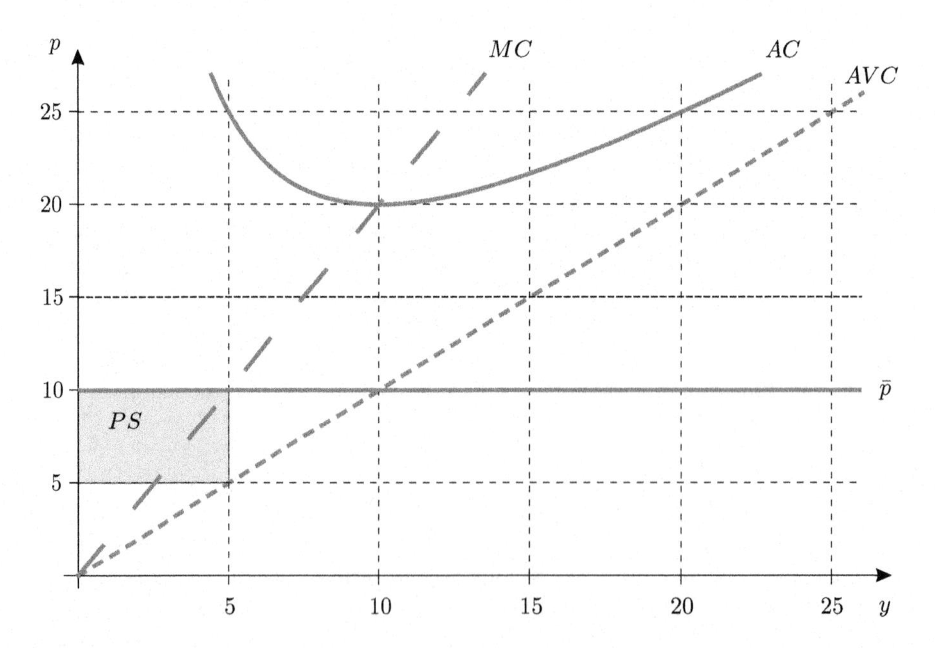

Figure 9.3 Exercise 1.1

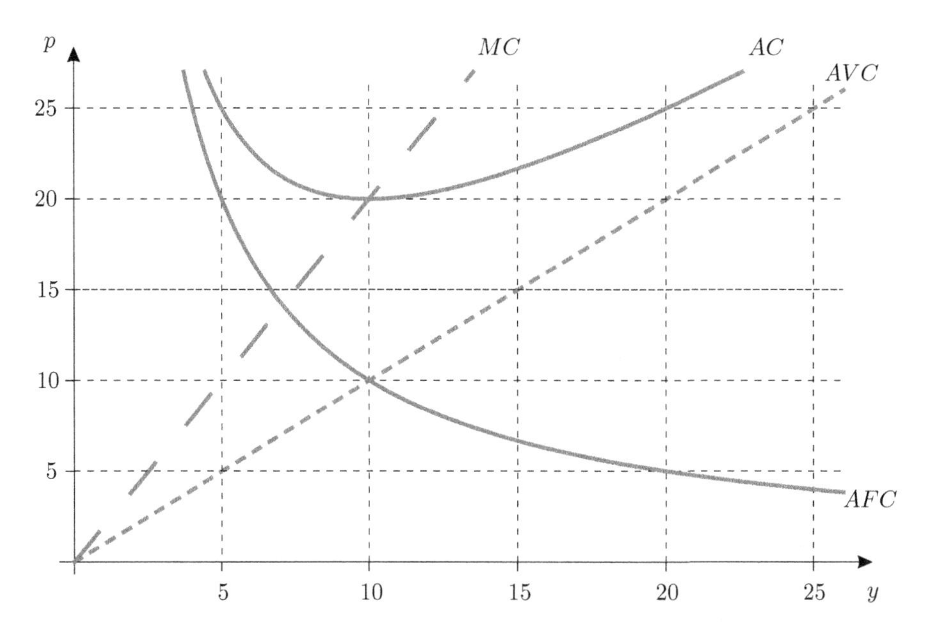

Figure 9.4 Exercise 1.2. AFC function

2. Because average variable costs $AVC(y)$ and average costs $AC(y)$ are not identical, fixed costs amount to $FC > 0$.[1] Average fixed costs are defined as:

$$AFC(y) = \frac{FC}{y}.$$

We know the following about this function:

$$\lim_{y \to 0} AFC(y) = \infty, \tag{9.1}$$

$$\lim_{y \to \infty} AFC(y) = 0, \tag{9.2}$$

$$AFC'(y) = -\frac{FC}{y^2}, \tag{9.3}$$

$$AFC''(y) = 2\frac{FC}{y^3}. \tag{9.4}$$

Thus, we know that $AFC(y)$ converges to infinity for small y (see Eq. 9.1), whereas it converges to zero for large y (see Eq. 9.2). Additionally, we know that it is a decreasing (see Eq. 9.3) and strictly convex (see Eq. 9.4) function.

[1] The graph alone gives no indication whether these fixed costs are technological (see Definition 8.6) or contractual fixed costs (see Definition 8.7). For this reason, we only know that $FC > 0$ if $y > 0$.

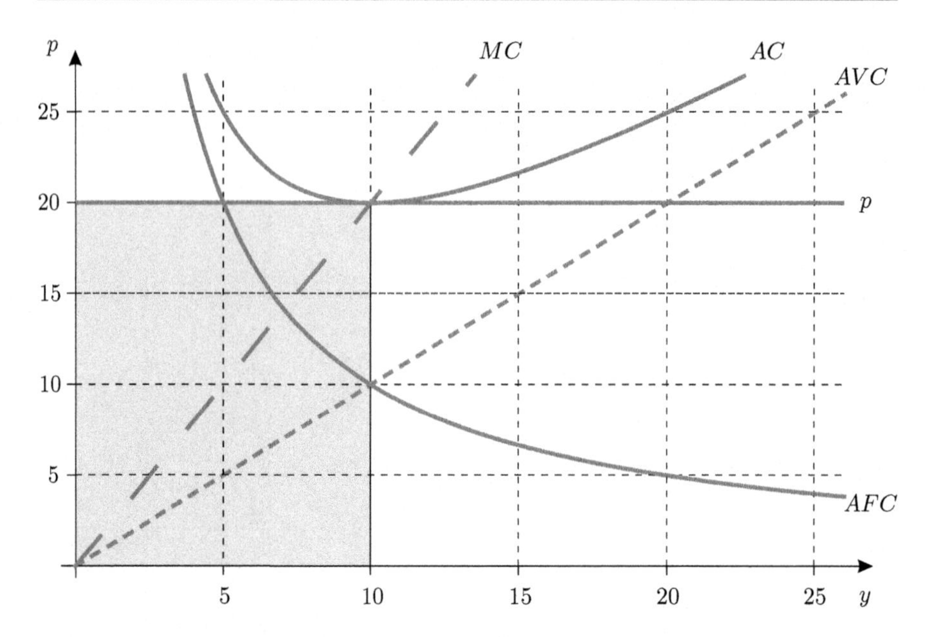

Figure 9.5 Exercise 1.3

Furthermore, we know the following (general) correlation:

$$AC(y) = AFC(y) + AVC(y) \quad \Leftrightarrow \quad AFC(y) = AC(y) - AVC(y). \quad (9.5)$$

We can see in Fig. 9.1 that the difference between AVC and AC at the point $y = 10$ is exactly 10, so that

$$AC(10) - AVC(10) = 10.$$

Thus, according to Eq. 9.5, $AFC(10) = 10$, and we get the following graph (see Fig. 9.4).

3. In the long-run equilibrium, all firms produce at the minimum of the average costs (AC). Here, the price is $p = 20$, and the amount provided by every firm is $y^* = 10$. Total costs correspond to $C(y^*) = y^* \cdot AC(y^*)$ (see the gray area in Fig. 9.5).
4. The cost function $C(y)$ consists of variable and fixed costs. There are two different ways to determine the variable costs (see Question 1): Figure 9.1 tells us that $AVC(y) = y$ and $MC(y) = 2\,y$, and thus $VC(y) = y^2$. We also know that $AFC(y) = \frac{FC}{y}$ and specifically $AFC(10) = 10$ (see Question 2). Which means that

$$AFC(10) = 10 \quad \Leftrightarrow \quad \frac{FC}{10} = 10 \quad \Leftrightarrow \quad FC = 100.$$

We thus get

$$C(y) = FC + VC(y) = 100 + y^2 \text{ for } y > 0. \tag{9.6}$$

5. The **long-run supply function** specifies the profit-maximizing amount y for every price $p \geq 0$. Obviously, there are fixed costs of $FC = 100$ for $y > 0$, which could be technological or contractual fixed costs. When determining the long-run supply function, we need to distinguish between the two cases.
   a) In the case of technological fixed costs, we know that:

$$\text{TFC} = \begin{cases} FC, & \text{for } y > 0, \\ 0, & \text{for } y = 0. \end{cases}$$

   For a firm's profit, it follows that

$$\Pi(y) = \begin{cases} p \cdot y - FC - VC(y) & \text{for } y > 0, \\ 0 & \text{for } y = 0. \end{cases} \tag{9.7}$$

   Thus, the firm will only produce according to the rule of optimization "price = marginal costs" if

$$\Pi(y^*) \geq 0 \Leftrightarrow p \cdot y^* - C(y^*) \geq 0 \Leftrightarrow p \geq \frac{C(y^*)}{y^*} \Leftrightarrow p \geq AC(y^*).$$

   This condition is only met if $p \geq AC_{\text{min}} = 20$, given that the monotonically increasing marginal-cost curve intersects with the average-cost curve at the minimum average cost. Hence, in the case of technological fixed costs, the long-term supply function is only identical to the inverse marginal-cost function if the price is at least as high as the minimum average costs AC_{min}. For a price below AC_{min}, the firm would make a negative profit if it supplied according to the optimization rule "price = marginal costs". Thus, the profit-maximizing firm avoids a loss by setting the supply (and, thus, also the profit) to zero (see Eq. 9.7). Utilizing the cost function as given by Eq. 9.6, the inverse marginal-cost function is:

$$MC^{-1}(p) = \frac{p}{2},$$

   and the long-run supply function is:

$$y(p) = \begin{cases} 0, & \text{for } p < 20, \\ \frac{p}{2}, & \text{else.} \end{cases}$$

b) In the case of contractual fixed costs we know that:

$$\text{CFC} = FC \text{ for } y \geq 0.$$

For a firm's profit, it follows that

$$\Pi(y) = p \cdot y - FC - VC(y),$$

for $y \geq 0$. This means that the firm produces according to the "price = marginal costs" rule once the price is large enough for the producer surplus $(PS(y^*))$ to be weakly positive, i.e. as soon as $p \geq AVC_{\min}$. As $AVC_{\min} = 0$, the inverse of the marginal-cost function represents a firm's long-run supply function for $p \geq 0$:

$$y(p) = \frac{p}{2} \text{ for } p \geq 0.$$

The **short-run supply curve** specifies the amount of y that maximizes producer surplus for every price $p \geq 0$. This means that the firm supplies according to the "price = marginal costs"-rule, once the price is large enough to cover the variable costs. Because we have a monotonically increasing marginal-cost function (with $MC(0) = 0$), this is the case for every positive price. Thus, the inverse of the marginal-cost curve represents the short-term supply function of the firm:

$$y(p) = \frac{p}{2} \text{ for } p \geq 0.$$

6. As labor is the only input, the costs are

$$C(y) = w \cdot L(y) = 2 \cdot L(y) \tag{9.8}$$

When we set Eq. 9.8 equal to the cost function given by Eq. 9.6, we find that[2]

$$100 + y^2 = 2 \cdot L(y)$$
$$\Leftrightarrow L^{-1}(l) = y = \sqrt{2l - 100}.$$

Thus, the production function is as follows:

$$Y(l) = \begin{cases} 0 & \text{for } l \leq 50, \\ \sqrt{2l - 100} & \text{else.} \end{cases}$$

For $l > 50$, we then receive a monotonically increasing and strictly concave function, as

- $Y'(l) = (2l - 100)^{-\frac{1}{2}}$ $(> 0 \text{ for } l > 50)$,
- $Y''(l) = -(2l - 100)^{-\frac{3}{2}}$ $(< 0 \text{ for } l > 50)$.

This function is illustrated in Fig. 9.6.

[2] The attentive reader will have noticed that there are two different terms whose square exactly corresponds to $2l - 100$. However, because we are looking only at positive outputs, the negative term $-\sqrt{2l - 100}$ can be omitted.

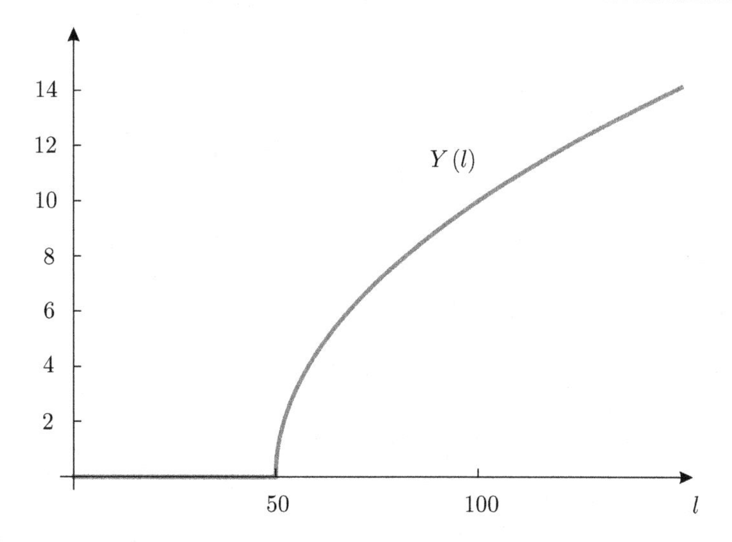

Figure 9.6 Exercise 1.6. Production function

9.2.2.2 Solutions to Exercise 2

1. It was assumed that the marginal-cost function increases linearly with output y. Thus,

$$MC = \beta + \alpha\, y, \tag{9.9}$$

with $\alpha > 0, \beta \geq 0$. If we integrate over y, then we receive the variable cost function $VC(y)$:

$$\int_0^y MC(z)\mathrm{d}z = VC(y) = \beta\, y + \frac{\alpha}{2}\, y^2.$$

It follows that

$$C(y) = FC + \beta \cdot y + \frac{\alpha}{2}\, y^2,$$

where $FC > 0$. Thus,

$$\begin{aligned} AC(y) &= \frac{FC}{y} + \beta + \frac{\alpha}{2}\, y, \\ AC'(y) &= -\frac{FC}{y^2} + \frac{\alpha}{2}, \\ AC''(y) &= \frac{2\, FC}{y^3}. \end{aligned}$$

Thus, we get a strictly convex average-cost function $AC(y)$.

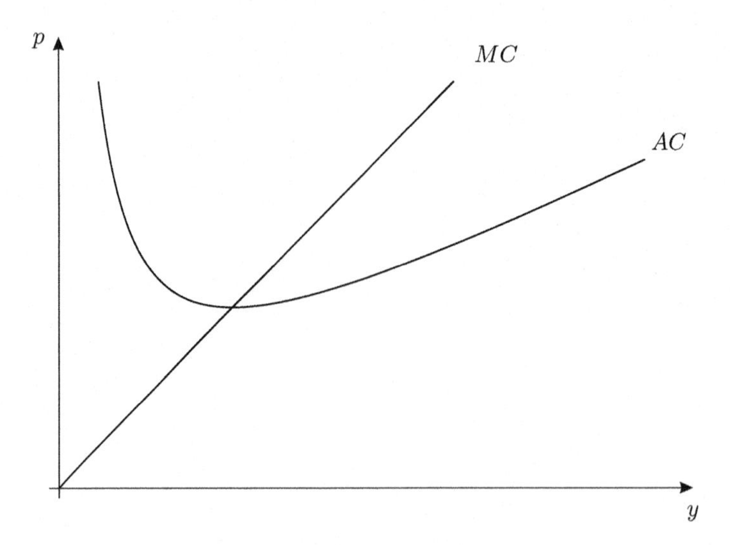

Figure 9.7 Exercise 2.1. Marginal- and average-cost function for $\beta = 0$

A marginal-cost curve and an average-cost curve for $\beta = 0$ are illustrated in Fig. 9.7. Take note that the marginal-cost function intersects the average-cost function in its minimum.
In the long-run equilibrium with free market entry and exit, the market price p^* corresponds to the minimum of the average costs AC_{min}. As the fixed costs are technological, Fig. 9.8 represents the supply function of booth i that supplies the amount of y_i^* at the market price p^*. The gray shaded area represents firm i's producer surplus at the optimum y_i^*. Since $\Pi_i(y_i^*) = 0$, the producer surplus corresponds to the fixed costs FC.
Market demand $x(p^*)$ corresponds to the market price p^*. In the long run, this demand is identical with the market supply of $n \cdot y_i^*$ (see Fig. 9.9). Thus, $n = 1{,}000$ sales-booths supply the identical quantity of y_i^* in equilibrium.
In the first question of this exercise we find the following:

- The profit of the sales booth is zero, because the market price p^* corresponds to the minimum of the average costs $AC_{\min}$.
- The producer surplus $PS(y^*)$ is positive and corresponds to the gray shaded area in Fig. 9.9.
- The consumer surplus $CS(y^*)$ is positive and corresponds to the triangle underneath the market demand function up until the equilibrium price p^* in Fig. 9.9.<?tex ?>

2. First of all, we have to check what kind of an impact the shortage of licenses has on the market. By reducing the licenses to $m = 800$, market supply is given by $m \cdot y_i^*$ under the old market price p^* , which is smaller than the corresponding market demand $x(p^*)$. Thus, the market price will increase until it reaches the new equilibrium $\tilde{p}^*$ (see Fig. 9.10). At this point, $x(\tilde{p}^*)$ equals supply of the 800 license owners ($m \cdot \tilde{y}_i^*$).

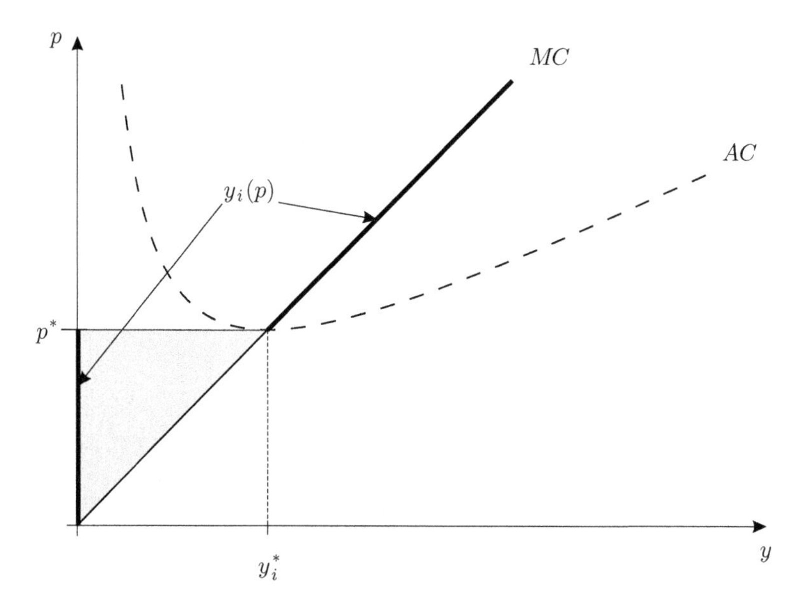

Figure 9.8 Exercise 2.1. Supply function of firm i

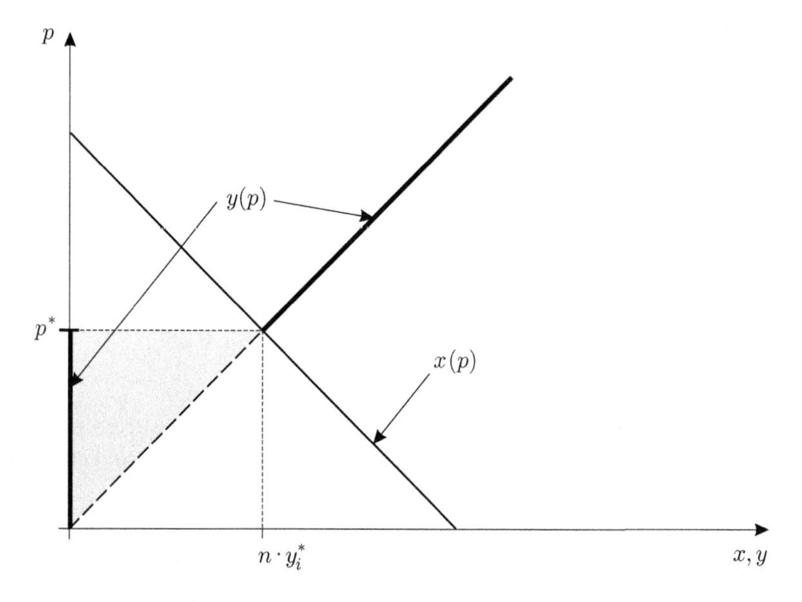

Figure 9.9 Exercise 2.1. Market supply function

In the sales booth i, which owns one of the 800 licenses, the optimal quantity goes up from y_i^* to $\tilde{y}_i^*$ because $\tilde{p}^* > p^*$ (see Fig. 9.11). Sales booth i makes a

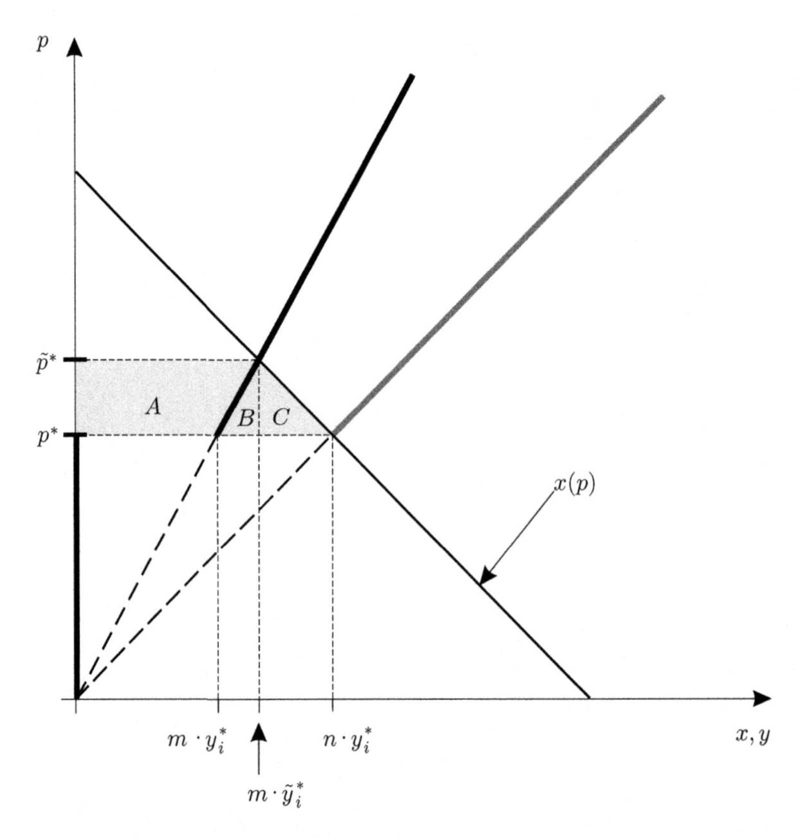

Figure 9.10 Exercise 2.2. New market equilibrium

positive profit because

$$\Pi_i(\tilde{y}_i^*) = \tilde{p}^* \cdot \tilde{y}_i^* - C(\tilde{y}_i^*) = \tilde{y}_i^* \left(\tilde{p}^* - AC(\tilde{y}_i^*)\right).$$

In the second question of this exercise we find the following:

- The equilibrium price rises through the shortage of licenses from p^* to $\tilde{p}^*$.
- The profits of the sales booths that own a license are positive and correspond to the gray area in Fig. 9.11.
- Not only the equilibrium quantity decreases ($m \cdot \tilde{y}_i^* < n \cdot y_i^*$), but also consumer surplus. The loss of consumer surplus corresponds to the gray area in Fig. 9.10:
  - Area A illustrates the profit of all 800 license holders and represents only a redistribution effect.
  - Area C is lost because some customers are not willing to pay the higher price $\tilde{p}^*$.
  - Area B is lost because all booths now produce at the equilibrium in the increasing part of the average-cost function.

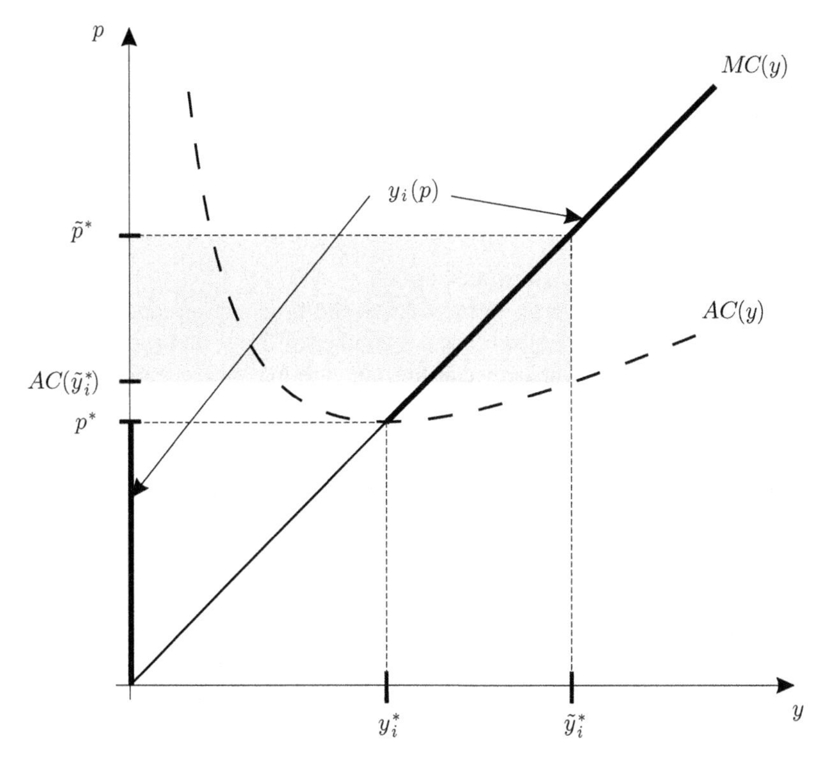

Figure 9.11 Exercise 2.2. Equilibrium supply of firm i

9.3 Multiple Choice

9.3.1 Problems

9.3.1.1 Exercise 1

Assume that firms are acting in a market with perfect competition and produce with identical cost functions $C(y) = y^2 + 16$. The market demand function is $x(p) = 200{,}008 - p$. Assume there is free market entry and exit.

1. Determine the equilibrium supply of a profit-maximizing firm.
   a) Indeterminate.
   b) 16.
   c) 4.
   d) 1.
   e) None of the above answers are correct.
2. Determine the price in the long-run equilibrium with free entry and exit.
   a) Indeterminate.
   b) 8.
   c) 1.

d) 4.
e) None of the above answers are correct.

3. Determine the number of firms supplying a positive quantity in the long-run equilibrium with free market entry and exit.
   a) Indeterminate.
   b) 50,000.
   c) 200,008.
   d) 200,000.
   e) None of the above answers are correct.
4. Assume that, due to a process innovation, the firms can produce with a cost function of $C(y) = 7y$. How does the number of firms, that supply a positive quantity, change in the long-run equilibrium with free market entry and exit?
   a) More firms are supplying the market.
   b) Fewer firms are supplying the market.
   c) The number of firms that are supplying the market is unchanged.
   d) No conclusion regarding the number of firms can be drawn.
   e) None of the above answers are correct.

9.3.1.2 Exercise 2

Assume that firms are acting in a market with perfect competition and produce with identical cost functions $C(y) = y$. The market demand function is $x(p) = 1{,}000{,}000 - p$. Assume that there is free market entry and exit.

1. Determine how many firms will offer a positive quantity in the long-run market equilibrium with free market entry and exit.
   a) Indeterminate.
   b) 1,000,000.
   c) 999,999.
   d) 0.
   e) None of the above answers are correct.

Now, assume that, in a market with perfect competition, firms operate with identical cost functions $C(y) = y^2 + 1$. The market demand function is $x(p) = 3{,}000{,}000 - p$. Assume that there is free market entry and exit.

2. Determine the supply of a profit-maximizing firm in the long-run market equilibrium with free market entry and exit.
   a) Indeterminate.
   b) 1.
   c) 2.
   d) 4.
   e) None of the above answers are correct.
3. Determine the market price in equilibrium.
   a) Indeterminate.
   b) 4.

c) 2.
d) 8.
e) None of the above answers are correct.

4. Determine how many firms will offer a positive quantity in the long-run equilibrium with free market entry and exit.
a) Indeterminate.
b) 3,000,000.
c) 750,000.
d) 2,999,998.
e) None of the above answers are correct.

9.3.1.3 Exercise 3

Assume that, in a market with perfect competition, 100 firms operate with an identical cost function $C(y) = y^2$. The market demand function is $x(p) = 1{,}000 - p$.

1. Determine the supply function of a profit-maximizing firm.
a) Indeterminate.
b) $y_i(p) = 2\,p$.
c) $y_i(p) = p$.
d) $y_i(p) = 0.5\,p$.
e) None of the above answers are correct.
2. Determine the market supply function of the profit-maximizing firms.
a) Indeterminate.
b) $y(p) = 200\,p$.
c) $y(p) = 100\,p$.
d) $y(p) = 50\,p$.
e) None of the above answers are correct.
3. Determine the equilibrium price in this market.
a) Indeterminate.
b) $\frac{1{,}000}{51}$.
c) $\frac{1{,}000}{101}$.
d) $\frac{1{,}000}{201}$.
e) None of the above answers are correct.
4. Determine the equilibrium quantity in this market.
a) Indeterminate.
b) $\frac{200{,}000}{201}$.
c) $\frac{100{,}000}{101}$.
d) $\frac{50{,}000}{51}$.
e) None of the above answers are correct.

9.3.1.4 Exercise 4

The market demand function for a certain good is given by $x(p) = \frac{b}{p}$ with $b > 0$. The long-run market supply function is $y(p) = p$.

1. Calculate the equilibrium quantity and price.
   a) The equilibrium price is $\sqrt{b}$ and the equilibrium quantity is $\sqrt{b}$.
   b) No equilibrium exists in this case.
   c) The equilibrium price is b^2 and the equilibrium quantity is b^2.
   d) The equilibrium price is b and the equilibrium quantity is b^2.
   e) None of the above answers are correct.
2. Assume the usual graphical representation, where the quantity is located on the abscissa and the price is on the ordinate. What does the short-run supply function's course look like compared to the long-run supply function's course?
   a) Where it is defined, the short-run supply function is above the long-run supply function.
   b) Where it is defined, the short-run supply function is below the long-run supply function.
   c) The short-term and long-run supply functions are congruent above the long-run average costs.
   d) Their position relative to each other cannot be determined with the given information.
   e) None of the above answers are correct.

9.3.1.5 Exercise 5

Assume that, in a market with perfect competition, firms produce with identical cost functions $C(y) = 0.5y^2 + 32$. Fixed costs are technological. The market demand function is $x(p) = 88 - p$.

1. Assume that there are 10 firms in the market and that neither market entry nor exit are possible in the short run. The cost and demand functions are as indicated above. Determine a profit-maximizing firm's supply function.
   a) $y(p) = p$ for $p \geq 8$ and $y(p) = 0$ for $p < 8$.
   b) $y(p) = p$.
   c) $y(p) = p$ for $p \geq y^2 + 1$ and $y(p) = 0$ for $p < y^2 + 1$.
   d) $y(p) = 0.5p$ for $p \geq y + \frac{1}{y}$ and $y(p) = 0$ for $p < y + \frac{1}{y}$
   e) None of the above answers are correct.
2. Determine the equilibrium price and quantity, given the same conditions as in Question 1.
   a) $p^* = 10$ and $x^* = y^* = 78$.
   b) $p^* = 44$ and $x^* = y^* = 44$.
   c) $p^* = 8$ and $x^* = y^* = 80$.
   d) The question cannot be answered with the given information.
   e) None of the above answers are correct.
3. Now assume there is free market entry and exit. Cost and demand functions are as indicated above. Determine how many firms offer a positive quantity in the long-run equilibrium with free market entry and exit.
   a) The number of firms is 10.
   b) The number of firms is 5.
   c) The number of firms is 0.

   d) The number of firms indeterminate.
   e) None of the above answers are correct.
4. Determine the equilibrium price in this market, given the same conditions as in Question 5.3.
   a) $p^* = 8$.
   b) $p^* = 88$.
   c) $p^* = 32$.
   d) With the given information the question cannot be answered.
   e) None of the above answers are correct.

9.3.1.6 Exercise 6

Two profit-maximizing firms who act as price-takers both have an inverse supply function $Q(y_i) = 200 + y_i$, where y_i is the quantity supplied by firm i and p is the market price. The demand function is $x(p) = 387.5 - \frac{p}{4}$.

1. Determine the inverse market supply function.
   a) The inverse market supply function is $P(y) = 400 + 2\,y$.
   b) The inverse market supply function is $P(y) = 200 + 2\,y$.
   c) The inverse market supply function is $P(y) = 200 + 0.5\,y$.
   d) The inverse market supply function is $P(y) = 400 + \;y$.
   e) None of the above answers are correct.
2. Determine the price and the quantity in equilibrium.
   a) The equilibrium price is $p^* = 300$, the quantity is $y^* = 400$.
   b) The equilibrium price is $p^* = 350$, the quantity is $y^* = 300$.
   c) The equilibrium price is $p^* = 400$, the quantity is $y^* = 200$.
   d) The equilibrium price is $p^* = 450$, the quantity is $y^* = 100$.
   e) None of the above answers are correct.
3. Determine the producer and consumer surplus.
   a) The consumer surplus is $CS(y^*) = 25{,}000$, the producer surplus is $PS(y^*) = 25{,}000$.
   b) The consumer surplus is $CS(y^*) = 180{,}000$, the producer surplus is $PS(y^*) = 22{,}500$.
   c) The consumer surplus is $CS(y^*) = 200{,}000$, the producer surplus is $PS(y^*) = 24{,}000$.
   d) The consumer surplus is $CS(y^*) = 92{,}500$, the producer surplus is $PS(y^*) = 28{,}000$.
   e) None of the above answers are correct.

A new firm with the same cost function enters the market. Everything else remains the same as before.

4. How do market quantity, price, and the total welfare in equilibrium change?
   a) The price increases, the quantity increases, welfare decreases.
   b) The price decreases, the quantity increases, welfare increases.
   c) The price decreases, the quantity decreases, welfare decreases.

d) The price decreases, the quantity increases, welfare decreases.
e) None of the above answers are correct.

9.3.1.7 Exercise 7

Assume that a firm supplies the international market for green coffee. The production of green coffee (y) requires labor (l). Additionally, there are technological fixed capital costs amounting to $TFC > 0$. The production function is given by $Y(l) = a\sqrt{l}$, with $a > 0$. The wage for a unit of labor is $w > 0$.

1. Determine the cost function $C(y)$ of the firm.
   a) Indeterminate.
   b) $C(y) = TFC + w\,a\,\sqrt{y}$.
   c) $C(y) = TFC + \frac{w\,a}{2\sqrt{y}}$.
   d) $C(y) = TFC + \frac{w}{a^2}\,y^2$.
   e) None of the above answers are correct.

Suppose $a = w = 2$ and $TFC = 72$.

2. How large is the profit-maximizing level of output y^* if the equilibrium price is given by $p^* = 40$? Determine the individual producer surplus $PS(y^*)$.
   a) Indeterminate.
   b) $y^* = 40$, $PS(y^*) = 800$.
   c) $y^* = 20$, $PS(y^*) = 600$.
   d) $y^* = \sqrt{40}$, $PS(y^*) = 250$.
   e) None of the above answers are correct.
3. The world market price for green coffee drops to $p^* = 12$ due to a decrease in demand. Determine the firm's profit-maximizing output y^* in the market. How large are profits $\pi(y^*)$ in the short run?
   a) Indeterminate.
   b) $y^* = 12$, $\pi(y^*) = 0$.
   c) $y^* = 6$, $\pi(y^*) = -18$.
   d) $y^* = \sqrt{12}$, $\pi(y^*) = 36$.
   e) None of the above answers are correct.
4. Suppose that all green coffee suppliers use the same production technology (see above). Total market demand is decreasing in the price, and the maximum willingness to pay for a unit of green coffee is 100. In an equilibrium with free market entry and exit, how high would the equilibrium price p^* be? Determine the level of output y^* supplied by a single firm in the equilibrium.
   a) Indeterminate.
   b) $p^* = 40$, $y^* = 40$.
   c) $p^* = \sqrt{12}$, $y^* = 24$.
   d) $p^* = 12$, $y^* = 12$.
   e) None of the above answers are correct.

9.3.1.8 Exercise 8

In a market with perfect competition, 2,000 consumers that have identical individual demand functions wish to consume a good. Individual demand is given by $x_i(p) = 1{,}000 - 0.0005 \cdot p$. Additionally, there are 1,000 suppliers with identical marginal costs functions, $MC(y_i) = 0.001 \cdot y_i - a$, where $a \leq 0$. There are no fixed costs.

1. Determine the market demand function $y(p)$ for this market.
   a) The market demand function is $x(p) = 2{,}000{,}000 - 0.0005 \cdot p$.
   b) The market demand function is $x(p) = 1{,}000 - p$.
   c) The market demand function is $x(p) = 2{,}000{,}000 - p$.
   d) The market demand function is $x(p) = 1{,}000 - 2 \cdot p$.
   e) None of the above answers are correct.
2. Determine the individual supply function $y_i(p)$ for this market.
   a) The individual supply function is $y_i(p) = 1{,}000 \cdot a + 1{,}000 \cdot p$.
   b) The individual supply function is $y_i(p) = 1{,}000 \cdot a + p$.
   c) The individual supply function is $y_i(p) = a + p$.
   d) The individual supply function cannot be determined with the given information.
   e) None of the above answers are correct.
3. Determine the market supply function $y(p)$ for this market.
   a) The market supply function is $y(p) = 1{,}000{,}000 \cdot a + 1{,}000{,}000 \cdot p$.
   b) The market supply function is $y(p) = 1{,}000{,}000 \cdot a + 1{,}000 \cdot p$.
   c) The market supply function is $y(p) = 1{,}000 \cdot a + p$.
   d) The market supply function cannot be determined with the given information.
   e) None of the above answers are correct.

Now, assume that the market demand function is given as $x(p) = 2{,}000 - 2\,p$ and the market supply function as $y(p) = 1{,}000\,b + 2\,p$, with $b \leq 0$.

4. Let $b = 0$. Determine the equilibrium price and quantity in this market.
   a) The equilibrium price is $p^* = 1{,}000$ and the equilibrium quantity is $x^* = y^* = 0$.
   b) The equilibrium price is $p^* = 0$ and the equilibrium quantity is $x^* = y^* = 2{,}000$.
   c) The equilibrium price is $p^* = 500$ and the equilibrium quantity is $x^* = y^* = 5{,}000$.
   d) The equilibrium price is $p^* = 500$ and the equilibrium quantity is $x^* = y^* = 1{,}000$.
   e) None of the above answers are correct.
5. Assume that a technological innovation decreases the marginal costs. How can this effect be illustrated in the model?
   a) Parameter b increases. The market supply function shifts to the right.
   b) Parameter b increases. The market supply function shifts upwards.
   c) Parameter b decreases. The market supply function shifts to the left.

d) Parameter b decreases. The market supply function shifts downwards.
e) None of the above answers are correct.

9.3.1.9 Exercise 9

Assume that the market demand for a good is given as $x(p) = 72 - p$ and that the technology on this market ensures an identical cost function with positive technological fixed costs for all firms, namely:

$$C(y_i) = \begin{cases} 36 + y_i^2, & \text{for } y_i > 0, \\ 0, & \text{for } y_i = 0. \end{cases}$$

The market is in the long-run equilibrium with free market entry and exit.

1. Determine a firm's long-run supply function $y_i(p)$ in this market.
   The supply function is:
   a)
   $$y_i(p) = \begin{cases} 0 & \text{for } p < 12, \\ \frac{p}{2} & \text{for } p \geq 12. \end{cases}$$
   b)
   $$y_i(p) = \begin{cases} 0 & \text{for } p < 6, \\ \frac{p}{2} & \text{for } p \geq 6. \end{cases}$$
   c)
   $$y_i(p) = \begin{cases} 0 & \text{for } p < 12, \\ 2 \cdot p & \text{for } p \geq 12. \end{cases}$$
   d)
   $$y_i(p) = \begin{cases} 0 & \text{for } p < 6, \\ 2 \cdot p & \text{for } p \geq 6. \end{cases}$$
   e) None of the above answers are correct.
2. Determine the total quantity supplied (y^*), the price (p^*) and the number of firms in the market (n^*) in the long-run equilibrium.
   a) $y^* = 20$, $p^* = 52$, $n^* = 8$.
   b) $y^* = 60$, $p^* = 12$, $n^* = 12$.
   c) $y^* = 40$, $p^* = 32$, $n^* = 10$.
   d) $y^* = 60$, $p^* = 12$, $n^* = 10$.
   e) None of the above answers are correct.
3. Determine the consumer surplus $CS(y^*)$ and the producer surplus $PS(y^*)$ in the equilibrium.
   a) $PS(y^*) = 100$, $CS(y^*) = 1{,}700$.
   b) $PS(y^*) = 360$, $CS(y^*) = 1{,}800$.
   c) $PS(y^*) = 500$, $CS(y^*) = 2{,}000$.

d) $PS(y^*) = 0, CS(y^*) = 1{,}700$.
e) None of the above answers are correct.

The state only hands out a finite number of permits for the supply of the good. Four firms, at most, are allowed to supply the good.

4. Determine the quantity y^* and the price p^* in the new equilibrium.
   a) $y^* = 40,\ p^* = 32$.
   b) $y^* = 48,\ p^* = 24$.
   c) $y^* = 60,\ p^* = 12$.
   d) $y^* = 32,\ p^* = 40$.
   e) None of the above answers are correct.
5. How does the sum of producer and consumer surplus change in comparison to Question 3?
   a) The sum of consumer and producer surplus increases.
   b) The sum of consumer and producer surplus does not change.
   c) The sum of consumer and producer surplus decreases.
   d) It is impossible to tell, because the identities of the firms remaining in the market are unknown.
   e) None of the above answers are correct.

9.3.2 Solutions

9.3.2.1 Solutions to Exercise 1

- Question 1, answer c) is correct.
- Question 2, answer b) is correct.
- Question 3, answer b) is correct.
- Question 4, answer d) is correct.

9.3.2.2 Solutions to Exercise 2

- Question 1, answer a) is correct.
- Question 2, answer b) is correct.
- Question 3, answer c) is correct.
- Question 4, answer d) is correct.

9.3.2.3 Solutions to Exercise 3

- Question 1, answer d) is correct.
- Question 2, answer d) is correct.
- Question 3, answer b) is correct.
- Question 4, answer d) is correct.

9.3.2.4 Solutions to Exercise 4

- Question 1, answer a) is correct.
- Question 2, answer c) is correct.

9.3.2.5 Solutions to Exercise 5

- Question 1, answer a) is correct.
- Question 2, answer c) is correct.
- Question 3, answer a) is correct.
- Question 4, answer a) is correct.

9.3.2.6 Solutions to Exercise 6

- Question 1, answer c) is correct.
- Question 2, answer b) is correct.
- Question 3, answer b) is correct.
- Question 4, answer b) is correct.

9.3.2.7 Solutions to Exercise 7

- Question 1, answer d) is correct.
- Question 2, answer b) is correct.
- Question 3, answer b) is correct.
- Question 4, answer d) is correct.

9.3.2.8 Solutions to Exercise 8

- Question 1, answer c) is correct.
- Question 2, answer a) is correct.
- Question 3, answer a) is correct.
- Question 4, answer d) is correct.
- Question 5, answer a) is correct.

9.3.2.9 Solutions to Exercise 9

- Question 1, answer a) is correct.
- Question 2, answer d) is correct.
- Question 3, answer b) is correct.
- Question 4, answer b) is correct.
- Question 5, answer c) is correct.

Firm Behavior in Monopolistic Markets 10

10.1 True or False

10.1.1 Statements

10.1.1.1 Block 1

1. The optimality condition “marginal costs = marginal revenues” characterizes the optimality condition only in a monopolistic but not in a perfectly competitive market.
2. Assume a non-price-discriminating monopolist who faces a decreasing demand function. Marginal revenues can be decomposed into a price and a quantity effect, and the price effect is always smaller than the quantity effect.
3. Assume a non-price-discriminating monopolist. Marginal revenues consist of a price and quantity effect. The price effect is always larger than the price effect under perfect competition.
4. If a firm owns a patent for a product, it can enforce prices above marginal costs, because the patent leads to a monopoly.

10.1.1.2 Block 2

Assume a profit-maximizing monopolist, who is selling an identical product in two different markets (third-degree price discrimination). He has constant marginal costs that are not equal to zero.

1. At the profit maximum, the firm will set a lower price in the market, in which the price-elasticity of demand in optimum (in absolute terms) is higher.
2. If demand is linear, the price-elasticity of demand can vary between zero and infinity (in absolute terms). The optimal pricing policy of the monopolist can therefore not be determined.
3. If price discrimination is prohibited between the two markets, it can be the case that the monopolist will cease to supply one of the markets.

M. Kolmar, M. Hoffmann, *Workbook for Principles of Microeconomics*,
Springer Texts in Business and Economics, https://doi.org/10.1007/978-3-319-62662-8_10

4. If price-discrimination becomes prohibited between the two markets, the price in the market with the higher price will always sink if the monopolist is still selling a positive quantity in both markets without price-discrimination.

10.1.1.3 Block 3

1. Assume a non-price-discriminating monopolist. His marginal revenues are positive, as long as the corresponding price is larger than zero.
2. Assume a non-price-discriminating monopolist. A profit-maximizing monopolist always supplies in the elastic part of the market demand function.
3. Assume a non-price-discriminating monopolist. In contrast to perfect competition, a profit-maximizing monopolist will make strictly positive profits with a constant marginal product.
4. A profit-maximizing monopolist, who is conducting perfect price discrimination, does not create any dead-weight loss.

10.1.1.4 Block 4

A profit-maximizing monopolist is confronted with an inverse market demand function $P(y) = 50 - \frac{y}{2}$. His cost function is composed of technological fixed costs TFC and variable costs $VC(y) = 2y^2$.

1. If the fixed costs are 200, then the monopolist will supply 10 units of the good.
2. If the fixed costs are 400, then the monopolist will supply 10 units of the good.
3. If the fixed costs are 200, then the monopolist will not supply anything.
4. If the fixed costs are 400, then the monopolist will not supply anything.

10.1.1.5 Block 5

A firm with a cost function $C(y) = 2y^2$ is confronted with an inverse market demand function $P(y) = 160 - 6y$. The supply function under perfect competition is identical to the monopolist's marginal-cost function.

1. The producer surplus with perfect competition and with symmetric firms is 768.
2. The producer surplus in a monopoly without price discrimination is 800.
3. The deadweight loss in a monopoly without price discrimination is 80.
4. The consumer surplus in a monopoly without price discrimination is 50.

10.1.1.6 Block 6

A monopolist with marginal costs of $MC(y) = d \cdot y$ is confronted with an inverse market-demand function $P(y) = 100 - y$, where $d > 0$, and she has zero fixed costs.

1. The non-price-discriminating monopolist will always supply the efficient quantity if $d = 0$.
2. Let $d = 2$. The equilibrium price of a non-price-discriminating monopolist is $p^* = 51$.

3. Let $d = 0$. With perfect price discrimination, the optimal profit of the monopolist is equal to 50.
4. The consumer surplus with perfect price discrimination is always strictly larger than the consumer surplus without price discrimination.

10.1.1.7 Block 7

1. Assume a non-price-discriminating monopolist. Her optimal supply of goods is always in the inelastic part of the market demand function.
2. Assume a price-discriminating monopolist, who produces with zero marginal costs and can discriminate prices between two different markets but not in the same market. The inverse market demand functions in both markets are $P(y_A) = a - y_A$ and $P(y_B) = b - y_B$. The profit-maximizing price in market A is always equal to the profit-maximizing price in market B.
3. Assume a non-price-discriminating monopolist, who produces with zero marginal costs. The inverse market demand function is $P(y) = \frac{1}{y}$. Then, all strictly positive quantities maximize the monopolist's profit.
4. Assume a price-discriminating monopolist, who produces with zero marginal and fixed costs and can discriminate prices between two different markets but not in the same market. The inverse market demand functions in both markets are $P(y_A) = a - y_A$ and $P(y_B) = b - y_B$. If $a = b > 0$, a prohibition of price discrimination has no effect on the behavior of the profit maximizing monopolist.

10.1.1.8 Block 8

1. A monopolist who can discriminate prices perfectly produces with marginal costs equal to zero. He is confronted with an inverse demand function of $P(y) = 100 - y$. The optimal producer surplus is $PS(y^*) = 5{,}000$.
2. A monopolist who cannot discriminate prices produces with marginal costs of zero. He is confronted with an inverse demand function of $P(y) = 100 - y$. In this case, the profit-maximizing quantity is $y^* = 25$.
3. In order for a monopolist to be willing to supply a positive quantity in a market, the producer surplus has to at least cover the fixed costs.
4. A monopolist who is performing third-degree price discrimination will supply the market with the larger price elasticity of demand in optimum (in absolute value) with a higher price.

10.1.1.9 Block 9

1. If a regulative authority prohibits third-degree price discrimination, consumer surplus will always increase.
2. If a regulative authority allows third-degree price discrimination, consumer surplus will always increase.

3. A health-insurance firm's franchise is an example of price discrimination.
4. If a profit-maximizing monopolist who can discriminate prices perfectly decides not to supply, then this is Pareto-efficient.

10.1.1.10 Block 10

1. A monopolist is confronted with a decreasing demand function. His or her marginal revenues are always lower than the price.
2. If the quantity effect dominates the price effect, the monopolist's revenues are always negative.
3. The difference between marginal revenues in a monopolistic market and marginal revenues in a perfectly competitive market is the existence of a (non-zero) price effect.
4. Monopolists as well as firms that are supplying under the conditions of perfect competition always apply the "marginal revenues = marginal costs" rule if they are supplying positive quantities.

10.1.1.11 Block 11

1. If *ComfyFoot* shoes are more expensive in Switzerland than in other countries, then this is an example of third-degree price discrimination.
2. The transition from a situation without to a situation with perfect price discrimination changes a Pareto-inefficient allocation into a Pareto-efficient one, but it need not be a Pareto-improvement.
3. Firms are interested in customer data because it allows them to discriminate prices more effectively.
4. A profit-maximizing monopolist who sells a positive quantity and engages in first-degree price discrimination calculates the optimal quantity according to the "marginal revenues = marginal costs" rule.

10.1.1.12 Block 12

1. A necessary condition for first-degree price discrimination is that the monopolist knows every potential buyer's willingness to pay.
2. With first-degree price discrimination, the producer surplus corresponds to the total welfare in the market.
3. A monopolist is confronted with two types of consumers who differ in their willingness to pay. In the monopolist's optimum, the monopolist offers contracts such that one of the group's consumer surplus is zero if he cannot identify the consumers.
4. Assume third-degree price discrimination and that the monopolist is a profit-maximizer. In the monopolist's optimum, one of the markets reacts more elastically than the other. In that case, the price is higher in the more elastic market.

10.1.2 Solutions

10.1.2.1 Sample Solutions for Block 1

1. **False**. The optimality condition "marginal costs = marginal revenues" holds also in a perfectly competitive market. See Chapter 9.1.
2. **True**. The profit maximum results from the monopolist's profit function $\pi(y) = P(y) \cdot y - C(y)$.

$$\overbrace{\underbrace{P'(y) \cdot y}_{=\text{ price effect}} + \underbrace{P(y)}_{=\text{ quantity effect}}}^{=\text{ marginal revenues}} - \overbrace{C'(y)}^{=\text{ marginal costs}} = 0.$$

   The monopolist's marginal profit is, thus, composed of a price and a quantity effect. Given a decreasing demand function, the price effect will be negative, whereas the quantity effect will be positive, for $y > 0$. Hence, the price effect is always smaller than the quantity effect. See Chapter 10.4.
3. **False**. When the monopolist is increasing his or her supply, the price effect is always negative, for $y > 0$. Correspondingly, there is no price effect in a perfectly competitive market, because all firms act as price takers.
4. **False**. If there is no demand for a good, then even a patent cannot ensure a (high) price for the good.

10.1.2.2 Sample Solutions for Block 2

1. **True**. The monopolist chooses the quantity that equalizes marginal revenues with marginal costs for each market. Given that the production of the two goods is identical, it follows that marginal costs are identical in both markets. From the explanation in Chapter 10.5.3, it follows that the good is sold at a lower price in the market with the higher price elasticity.
2. **False**. Consider the derivation of a monopolist's optimal pricing strategy for a linear demand function in Chapter 10.4.
3. **True**. Say that market 2 is the market with the lower price elasticity of demand. If the difference in the marginal willingness to pay is large enough, then the monopolist might set a price so high that only market 2 is served. See Chapter 10.5.3.
4. **True**. In this case, the monopolist will offer the good in both markets at a price that is between the two different prices with price discrimination. See Chapter 10.5.3.

10.1.2.3 Sample Solutions for Block 3

1. **False**. Marginal revenues are negative if the absolute value of the price effect is larger than that of the quantity effect. See Chapter 10.4.

2. **True**. The fact that the optimal quantity and price are in the elastic part of the demand function is no coincidence: if demand is inelastic, then the monopolist can increase revenues by reducing output, because a one percent decrease in output increases the price by more than one percent. However, in this case, the initial level of output could not have been profit-maximizing, since reducing the output also reduces costs. See Chapter 10.4.
3. **True**. See Chapter 10.4.
4. **True**. The outcome in the market under perfect price discrimination is Pareto-efficient. Contrary to perfect competition, however, the entire welfare goes to the monopolist in the form of producer surplus. See Chapter 10.5.1.

10.1.2.4 Sample Solutions for Block 4

The monopolist's revenues are $R(y) = P(y) \cdot y = (50 - 0.5y)\, y = 50\, y - 0.5\, y^2$. Thus, marginal revenues are $MR(y) = R'(y) = 50 - y$. The marginal costs are $MC(y) = C'(y) = 4\, y$. In the optimum, "marginal revenues = marginal costs" holds, i.e.:

$$\begin{aligned} MR(y) &= MC(y) \\ \Leftrightarrow 50 - y &= 4\, y \\ \Leftrightarrow \quad y^* &= 10. \end{aligned}$$

Inserting the optimal quantity into the inverse demand function gives the price at the optimum:

$$\begin{aligned} p^* &= P(y^*) \\ \Leftrightarrow p^* &= 50 - 0.5\, y^* = 45. \end{aligned}$$

However, the monopolist only supplies the good if he does not incur losses, i.e.:

$$\begin{aligned} \pi(y^*) &\geq 0 \\ \Leftrightarrow \quad y^* \cdot p^* - C(y^*) = 10 \cdot 45 - (2 \cdot 10^2 + TFC) &\geq 0 \\ \Leftrightarrow \quad 250 &\geq TFC. \end{aligned}$$

Thus, the monopolist's optimal supply is

$$y^* = \begin{cases} 10 & \text{for } TFC \leq 250, \\ 0 & \text{else.} \end{cases}$$

It follows for the four statements that

1. **True**.
2. **False**.
3. **False**.
4. **True**.

10.1.2.5 Sample Solutions for Block 5

- In an equilibrium with perfect competition we have $x = y$ and

$$\begin{aligned} P(x) &= MC(y) \\ \Leftrightarrow 160 - 6x &= 4y \\ \Leftrightarrow \quad y^* &= 16. \end{aligned}$$

It follows that $p^* = 64$ and

$$PS(y^*) = 0.5 \cdot y^* \cdot p^* = 512$$

and

$$CS(y^*) = 0.5 \cdot (P(0) - p^*) \cdot y^* = 0.5 \cdot (160 - 64) \cdot 16 = 768.$$

- The following holds true in a monopoly:

$$\begin{aligned} MR(y) &= MC(y) \\ \Leftrightarrow 160 - 12y &= 4y \\ \Leftrightarrow \quad y^* &= 10. \end{aligned}$$

It follows that $p^* = 100$ and, given that there are no fixed costs,

$$PS(y^*) = \pi(y^*) = p^* \cdot y^* - C(y^*) = 10 \cdot 100 - 2 \cdot 10^2 = 800$$

and

$$CS(y^*) = 0.5 \cdot (P(0) - p^*) \cdot y^* = 0.5 \cdot (160 - 100) \cdot 10 = 300.$$

The deadweight loss is therefore

$$DWL = 512 + 768 - (300 + 800) = 180.$$

It follows for the four statements that

1. **False**.
2. **True**.
3. **False**.
4. **False**.

10.1.2.6 Sample Solutions for Block 6

1. **False**. In the optimum, the monopolist chooses her quantity according to the "marginal revenues = marginal costs" rule, i.e.

$$MR(y) = 100 - 2y = 0 = MC(y)$$
$$\Leftrightarrow \quad y^* = 50.$$

However, if $d = 0$, the efficient quantity is 100.
2. **False**. This would be true for $MC(y) = d$. For $MC(y) = d\, y = 2\, y$, however, we get

$$MR(y) = MC(y)$$
$$\Leftrightarrow 100 - 2y = 2y$$
$$\Leftrightarrow y^* = 25$$

It follows that $P(y^*) = 100 - y^* = 75$.
3. **False**. If fixed costs are zero, producer surplus is equivalent to profits. With perfect price discrimination and marginal costs of zero, the producer surplus corresponds to the entire area underneath the demand function, i.e., $PS(y^*) = \pi(y^*) = \frac{1}{2} \cdot 100 \cdot 100 = 5{,}000$.
4. **False**. With perfect price discrimination, the consumer surplus is zero, whereas it is strictly positive without price discrimination. See Chapter 10.5.

10.1.2.7 Sample Solutions for Block 7

1. **False**. See Chapter 10.4.
2. **False**. The optimality conditions for third-degree price discrimination are:

$$\frac{\partial P_A(y_A)}{\partial y_A} \cdot y_A + P_A(y_A) = \frac{\partial C(y_A + y_B)}{\partial y_A}$$

and

$$\frac{\partial P_B(y_B)}{\partial y_B} \cdot y_B + P_B(y_B) = \frac{\partial C(y_A + y_B)}{\partial y_B}.$$

As marginal costs are zero, it follows that

$$\frac{\partial C(y_A + y_B)}{\partial y_A} = 0 = -1 \cdot y_A + a - y_A$$

and

$$\frac{\partial C(y_A + y_B)}{\partial y_B} = 0 = -1 \cdot y_B + b - y_B.$$

Thus, $y_A^* = \frac{a}{2}$, $p_A^* = a - \frac{a}{2} = \frac{a}{2}$, $y_B^* = \frac{b}{2}$, $p_B^* = b - \frac{b}{2} = \frac{b}{2}$. Therefore, the price depends on a and b and thus would only be identical if $a = b$. See Chapter 10.5.3.

3. **True**. The optimality condition without price discrimination is:

$$P'(y) \cdot y + P(y) = C'(y).$$

Because marginal costs are zero, it follows that

$$-y^{-2} \cdot y + y^{-1} = 0 \Leftrightarrow -y^{-1} + y^{-1} = 0$$

which is true for all values of $y > 0$. See Chapter 10.4.

4. **True**. In this case the optimal price in case of price discrimination ($p_A^{PD} = \frac{a}{2} = p_B^{PD} = \frac{b}{2}$) is identical to the optimal price in case price discrimination is prohibited ($p^* = \frac{a}{2} = \frac{b}{2}$).

10.1.2.8 Sample Solutions for Block 8

1. **True**. The producer surplus is revenues minus variable costs. In this case, variable costs are zero and revenues are equal to the area below the demand function, i.e. $100 \cdot 100 \cdot \frac{1}{2} = 5{,}000$. See Chapter 10.5.1.
2. **False**. The optimality condition "marginal revenues = marginal costs" leads to:

$$\begin{aligned} & P'(y) \cdot y + P(y) = C'(y) \\ \Leftrightarrow \quad & -y + (100 - y) = 0 \\ \Leftrightarrow \quad & y^* = 50. \end{aligned}$$

See Chapter 10.4.

3. **False**. In order for a monopolist to be willing to supply a positive quantity, the producer surplus has to at least cover the variable costs. See Chapter 10.4.
4. **False**. A monopolist who is able to conduct third-degree price discrimination will charge a higher price in the market with the lower price elasticity of demand (in absolute value). See Chapter 10.5.3.

10.1.2.9 Sample Solutions for Block 9

1. **False**. If it is more profitable for the monopolist to only supply the market with the lower price elasticity, consumer surplus will decrease. See Chapter 10.5.3.
2. **False**. This statement is only true if the monopolist does not supply both markets when price discrimination is prohibited. See Chapter 10.5.3.
3. **True**. The franchise is the amount the patient has to pay out of her own pocket if she becomes ill. Each policyholder can choose how high this amount shall be, which in turn influences the insurance costs (the premium). (This is not to be confused with the deductible, which has to be paid proportionally to the

costs when costs exceed the franchise.) Thus, different units of the product "health insurance" are sold at different prices. (Please note: the deductible is not only dependent on the willingness to pay, it is also largely regulated by the government.)

4. **True**. A monopolist who conducts perfect price discrimination will supply a positive quantity as long as the price at least covers marginal costs. If she is not supplying the good, then the willingness to pay of the customer with the highest willingness to pay is below the marginal costs. See Chapter 10.5.1.

10.1.2.10 Sample Solutions for Block 10

1. **False**. Counterexample: The monopolist's marginal revenues are

$$R'(y) = P'(y) \cdot y + P(y),$$
$$R'(0) = P(0).$$

2. **False**. His revenues will always be positive.
3. **True**. From the monopolist's profit function $\pi(y) = P(y) \cdot y - C(y)$ it follows for the profit maximum:

$$\overbrace{\underbrace{P'(y) \cdot y}_{= \text{price effect}} + \underbrace{P(y)}_{= \text{quantity effect}}}^{= \text{marginal revenues}} - \overbrace{C'(y)}^{= \text{marginal cost}} = 0.$$

A monopolist's marginal revenues can be decomposed into a price effect and a quantity effect. See Chapter 10.4.

4. **True**. The condition "marginal revenues = marginal costs" is always optimal, independent of the market structure. See Chapter 10.4 and Chapter 9.3.

10.1.2.11 Sample Solutions for Block 11

1. **True**. If an identical good is offered in different markets for two different prices, it is a typical case of third-degree price discrimination. See Chapter 10.5.3.
2. **True**. In a situation without perfect price discrimination, there is always a deadweight loss (see for example Figure 10.1 in Chapter 10.4); whereas a situation with perfect price discrimination does not constitute a deadweight loss (see Figure 10.2 in Chapter 10.5.1). Allowing for perfect price discrimination thus changes an initially Pareto-inefficient allocation into a Pareto-efficient one. However, referring again to the same figures, one can see that the re-allocation leads to a full decline of consumer surplus, which implies that the consumers are worse off. Thus, there is no Pareto improvement. See the definition of Pareto improvement in Chapter 5.1.
3. **True**. Based on customer data, firms can better determine the customers' willingness to pay, which often allows for more effective price discrimination.
4. **True**. The monopolist profit increases as long as the price of the last unit exceeds the marginal costs of that unit. Hence, she will expand her supply up until the point where price equals marginal costs. See Chapter 10.5.1.

10.1.2.12 Sample solutions for Block 12

1. **True**. It is a necessary but not a sufficient condition (if, for example, the monopolist cannot discriminate prices because of a prohibition of perfect price discrimination). See Chapter 10.5.1.
2. **True**. With first-degree price discrimination, there is no deadweight loss, because the monopolist converts the entire consumer surplus into producer surplus. As a result, total welfare in the market corresponds to the producer surplus. See Chapter 10.5.1.
3. **True**. The L-type's consumer surplus is zero and the H-type's is positive (unless the L-type is not supplied at all). See Chapter 10.5.2.
4. **False**. The product is sold for a higher price in the market which has a less elastic demand function. See Chapter 10.5.2.

10.2 Open Questions

10.2.1 Problems

10.2.1.1 Exercise 1

The (inverse) demand on a market is given by $P(y) = 300 - \frac{1}{2}\,y$.

1. Assume that the seller is a monopolist, who cannot discriminate prices.
   a) How large is the monopolist's revenue at an output of $y = 200$ and how large is it at $y = 400$?
   b) How large are the marginal revenues in both cases if one raises the output quantity by $\Delta\, y = 50$? How large is the quantity effect (QE) and how large is the price effect (PE)?
   c) Plot the monopolist's revenue function and calculate the revenues at the revenue maximum.
   d) How large is the price elasticity of demand at the revenue-maximum?
   e) Assume that the firm's fixed costs and the firm's marginal costs are zero. How large is the monopolist's supply at the profit maximum, and how large is the price elasticity of demand? How do the results change if we assume constant marginal costs of $0 < c < 300$?
2. Assume that the non-price-discriminating monopolist has the following cost function:

$$C(y) = \begin{cases} 0, & \text{for } y = 0, \\ TFC + 100 \cdot y, & \text{for } y > 0, \end{cases}$$

   with $TFC > 0$.
   a) Calculate the profit-maximizing supply (y^*) contingent on TFC. How large can TFC be at most to make sure that $y^* > 0$?

Assume that $TFC = 10{,}000$.

(b) Calculate the profit-maximizing supply (y^*) and price (p^*).

(c) Calculate the sum of the consumer surplus (CS) and the producer surplus (PS) as well as the deadweight loss (DWL).

10.2.1.2 Exercise 2

A monopolist faces two different consumer types: The H-type has a high willingness to pay (WTP) for the good supplied by the monopolist, whereas the L-type has a low WTP. Their marginal WTP are given by

$$P_L(x) = 5 - x \text{ and } P_H(x) = 8 - x.$$

Assume there is only one individual of each type. Costs of production are zero, i.e. $MC = FC = 0$.

1. For each type, determine the WTP, $W_j(x)$, with $j \in \{L, H\}$.
2. The monopolist can identify both customers as either L-type or H-type and has all the relevant information about their respective WTP. Price discrimination is allowed. Determine the profit-maximizing quantities x_L^* and x_H^*. What is the monopolist's profit, and what is the consumer surplus? Is the allocation Pareto-efficient?
3. The monopolist has complete information about the WTP. However, he cannot identify a customer as either L-type or H-type. Price discrimination is allowed.
   a) Assume the monopolist offers a single contract that is accepted by the H-type. What is the quantity x and the price p that is specified in this contract $\{x, p\}$ if the monopolist maximizes his profits? What is the monopolist's profit in optimum, and what is the consumer surplus for this contract?
   b) Assume the monopolist offers a single contract which is accepted by the L-type. What is the quantity x and the price p in this contract $\{x, p\}$ if the monopolist maximizes his profits? What is the monopolist's profit, and what is the consumer surplus?
   c) Assume the monopolists offers two contracts: one for the H-type and one for the L-Type. Further assume that the monopolist supplies the Pareto-efficient quantity. What is the price in each contract, $\{x_L^*, p_L\}$ and $\{x_H^*, p_H\}$ if the monopolist maximizes his profits? What is the monopolist's profit, and what is the consumer surplus?
   d) Assume that the monopolist supplies the Pareto-efficient quantity x_H^* for the H-type, but lowers his supply for the L-type to $\tilde{x}_L = 2$. How do the optimal prices change in the two contracts compared to Question 3c)? What can you say about the profits and consumer surplus? Was it a good decision for the monopolist to adjust his supply?
   e) Determine the profit-maximizing prices p_L and p_H, as well as the profit-maximizing quantity. What is the monopolist's profit, what is the consumer surplus?

10.2.1.3 Exercise 3

The demand, on two different markets for a single good, are given by:

$$x_1(p_1) = 300 - 2\,p_1 \text{ and } x_2(p_2) = 200 - 2\,p_2.$$

The marginal costs of production of a price discriminating monopolist (third-degree price discrimination) are given by $c \geq 0$.

1. Determine the monopolist's optimal price and quantity.
2. Under which conditions will the monopolist supply both markets?
3. Determine the price and the quantities on both markets for $c = 50$. Determine the price elasticity of demand on both markets at the optimum. Explain the relationship between the price elasticity of demand and the price on a market.
4. Determine the monopolist's profit function assuming that price discrimination is prohibted. Illustrate the (aggregate) demand function with the help of a graph.
5. Determine the optimal monopoly price and monopoly quantity assuming that price discrimination is prohibited. Under which conditions will the monopolist only supply the first market?

10.2.2 Solutions

10.2.2.1 Solutions to Exercise 1

1. We assume a non-price-discriminating monopolist
   a) Revenues are equal to price times quantity. In the case of a monopolist, the price is a function of quantity. This means, the higher the quantity that he wants to sell, the lower the price that he can charge. In general, the following holds for a monopolist's revenues:

   $$R(y) = P(y)\,y.$$

   Here, for $y = 200$ and $y = 400$ we get

   $$R(200) = \underbrace{\left(300 - \frac{1}{2} \cdot 200\right)}_{=P(y)} \cdot \underbrace{200}_{=y} = 200 \cdot 200 = 40{,}000,$$

   $$R(400) = \underbrace{\left(300 - \frac{1}{2} \cdot 400\right)}_{=P(y)} \cdot \underbrace{400}_{=y} = 100 \cdot 400 = 40{,}000.$$

   The monopolist's revenues are identical in both cases.

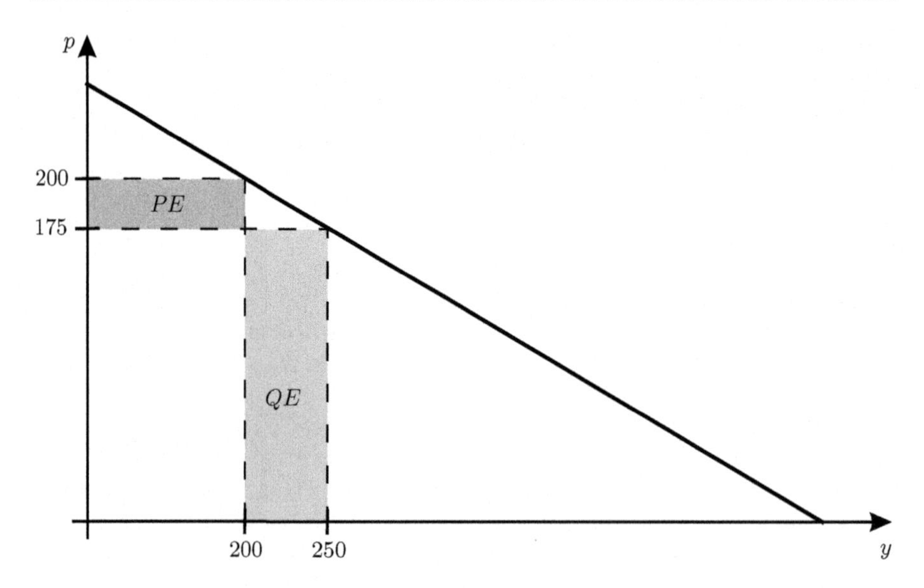

Figure 10.1 Exercise 1.1a). Price- and Quantity Effect

b) A change in quantity of $\Delta y = 50$ also changes the revenues

$$R(250) = \left(300 - \frac{1}{2} \cdot 250\right) \cdot 250 = 175 \cdot 250 = 43{,}750,$$

$$R(450) = \left(300 - \frac{1}{2} \cdot 450\right) \cdot 450 = 75 \cdot 450 = 33{,}750.$$

The change in revenues ΔR (i.e., the marginal revenue) is

$$\Delta R = R(250) - R(200) = 3{,}750, \tag{10.1}$$

$$\Delta R = R(450) - R(400) = -6{,}250. \tag{10.2}$$

Starting from $y = 200$, the monopolist's revenues increase when output is increased by $\Delta y = 50$; while it decreases if you start from $y = 400$.
We can see the price/quantity-combination before and after the change in quantity in the case of $y = 200$ in Fig. 10.1.

- PE: The price effect corresponds to the change in revenues due to the reduced price. Before the change, the monopolist sold $y = 200$ units for a price of $P(200) = 200$ per unit. After the change, the price is $P(250) = 175$. This means that he now sells the 200 units (that he sold for 200 Swiss Francs before) for 175 Swiss Francs. The (negative) price effect thus corresponds to

$$PE = y \cdot (P(y + \Delta y) - P(y)) = 200 \cdot (175 - 200) = -5{,}000.$$

- QE: The quantity effect corresponds to the change in revenues due to the increase in supply. The monopolist sold $y = 200$ units before the change, and sells $y = 250$ units afterwards. This means, he can sell 50 additional units if the price is reduced to $P(250) = 175$. The (positive) quantity effect corresponds to

$$QE = P(y + \Delta y) \cdot \Delta y = 175 \cdot 50 = 8{,}750.$$

We can see that, starting from $y = 200$, the (positive) quantity effect of an increase in y overcompensates the (negative) price effect. Hence, the total effect is positive (see also Eq. 10.1):

$$\Delta R = PE + QE = -5{,}000 + 8{,}750 = 3{,}750.$$

The results change once we change the starting point to $y = 400$. While the absolute value of the (negative) price effect increases in comparison to $y = 200$,

$$PE = y \cdot (P(y + \Delta y) - P(y)) = 400 \cdot (75 - 100) = -10{,}000,$$

the (positive) quantity effect decreases

$$QE = P(y + \Delta y) \cdot \Delta y = 75 \cdot 50 = 3{,}750.$$

This causes the total effect on marginal revenues to become negative (see also Eq. 10.2):

$$\Delta R = PE + QE = -10{,}000 + 3{,}750 = -6{,}250.$$

c) In order to plot the revenue function, we first calculate the first two derivatives of $R(y) = 300\,y - \frac{1}{2}\,y^2$:

$$R'(y) = 300 - y,$$
$$R''(y) = -1.$$

Hence, we have a strictly concave function that has its maximum at the point

$$R'(y) = 0 \quad \Leftrightarrow \quad \hat{y} = 300.$$

The price at the revenue-maximum is $\hat{p} = P(\hat{y}) = 150$. Revenues at the revenue-maximum are therefore

$$R(\hat{y}) = \hat{p} \cdot \hat{y} = 150 \cdot 300 = 45{,}000.$$

Figure 10.2 illustrates the results. Marginal revenues (the slope of the revenue-function) are positive for $y < \hat{y}$. The positive quantity effect overcompensates for the negative price effect.
Marginal revenues are negative for $y > \hat{y}$: In this case, the negative price effect overcompensates for the positive quantity effect.

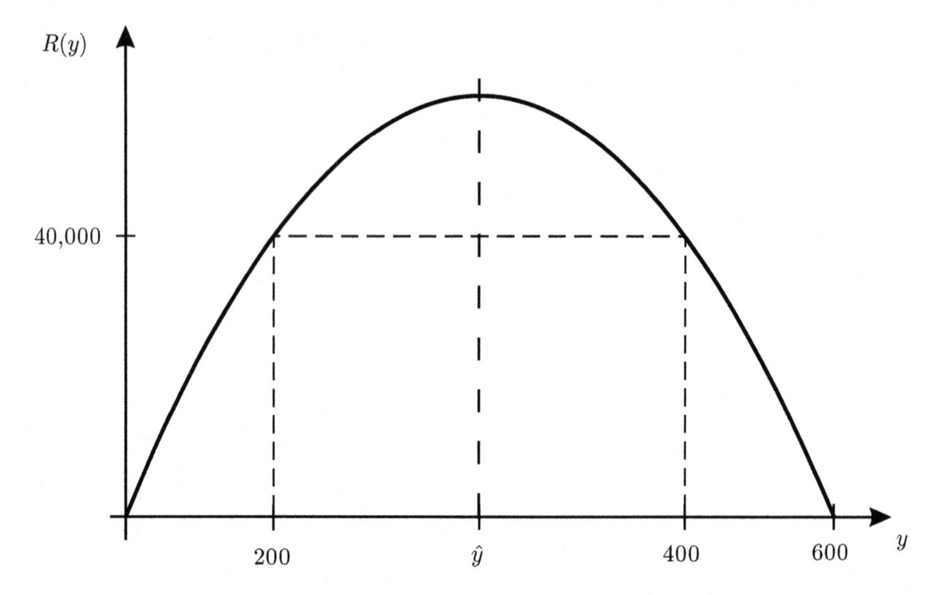

Figure 10.2 Exercise 1.1c). The Revenue Function

d) The price elasticity of demand is

$$\epsilon_p^x = \frac{d\,x(p)}{d\,p}\frac{p}{x(p)}.$$

The demand function $x(p)$ is computed by using the inverse demand function (note that $x = y$):

$$P(y) = 300 - \frac{1}{2}\,y \quad \Leftrightarrow \quad x(p) = 600 - 2\,p.$$

We get

$$\epsilon_{\hat{p}}^x = -2 \cdot \frac{\hat{p}}{600 - 2\,\hat{p}} = -2 \cdot \frac{150}{600 - 2 \cdot 150} = -1.$$

Hence, the demand is isoelastic at the revenue-maximum.

e) If marginal and fixed costs are equal to zero, the problem of profit maximization is identical to the problem of revenue maximization since $\pi(y) = R(y)$ in this case. Thus, we get $y^* = \hat{y} = 300$, $p^* = \hat{p} = 150$ and $\pi^* = \pi(y^*) = R(\hat{y}) = 45{,}000$. The price elasticity of demand at p^* is $\epsilon_{p^*}^x = \epsilon_{\hat{p}}^x = -1$.
Next, we determine the profit maximum for zero fixed costs and marginal costs of $0 < c < 300$. In this case, the firm's profit-function is

$$\pi(y) = \left(300 - \frac{1}{2}\,y\right)y - c\,y.$$

And at the profit maximum:

$$\pi'(y) = 0 \Leftrightarrow \underbrace{300 - y}_{= \text{marginal revenues}} - \underbrace{c}_{= \text{marginal costs}} = 0 \Leftrightarrow y = 300 - c. \quad (10.3)$$

Thus, the profit-maximizing quantity is $y^* = 300 - c$, and the price is $p^* = P(y^*) = 300 - \frac{1}{2} \cdot (300 - c) = 150 + \frac{c}{2}$. The higher the marginal costs, the lower the monopoly quantity at the optimum, and the higher the monopoly price.

What is the price elasticity of the demand at the optimum? We know that

$$\epsilon_{p^*}^{y} = -2 \cdot \frac{p^*}{y(p^*)} = -2 \cdot \frac{150 + \frac{c}{2}}{300 - c} = -\frac{300 + c}{300 - c} < -1,$$

where the latter inequality holds since the numerator is always larger than the denominator for $c > 0$. Thus, the monopolist's supply always lies in the elastic part of the demand function for $c > 0$.

2. There are technological fixed costs of $TFC > 0$.
   a) Because technological fixed costs only show up in case of production ($y > 0$), we have to check whether revenues at least cover total costs (variable costs plus technological fixed costs). If this were not the case, the monopolist would have losses, which he would avoid by withdrawing from the market. We first determine the optimal quantity of the monopolist under the assumption that TFC is small enough. Utilizing Eq. 10.3, this quantity is given by $y^* = 300 - c = 300 - 100 = 200$ and the monopoly price is $p^* = 150 + \frac{c}{2} = 200$. The monopoly profit corresponds to

$$\begin{aligned} \pi^* = \pi(y^*) &= \underbrace{p^* \cdot y^*}_{= \text{revenues}} - \underbrace{(TFC + c \cdot y^*)}_{= \text{costs}} \\ &= 40{,}000 - (TFC + 20{,}000) \\ &= 20{,}000 - TFC. \end{aligned}$$

   We can now figure out the critical upper limit of TFC. If $\pi(y^*) \geq 0$, we get

$$\pi^* \geq 0 \quad \Leftrightarrow \quad 20{,}000 - TFC \geq 0 \quad \Leftrightarrow \quad 20{,}000 \geq TFC.$$

   Thus, profit-maximizing output contingent on TFC is given by

$$y^*(TFC) = \begin{cases} 0, & \text{for } TFC > 20{,}000, \\ 200, & \text{for } TFC \leq 20{,}000. \end{cases}$$

   This result is illustrated in Fig. 10.3.
   b) Assuming that $TFC = 10{,}000$, we find that $y^* = 200$, $p^* = 200$, and $\pi^* = 10{,}000$.

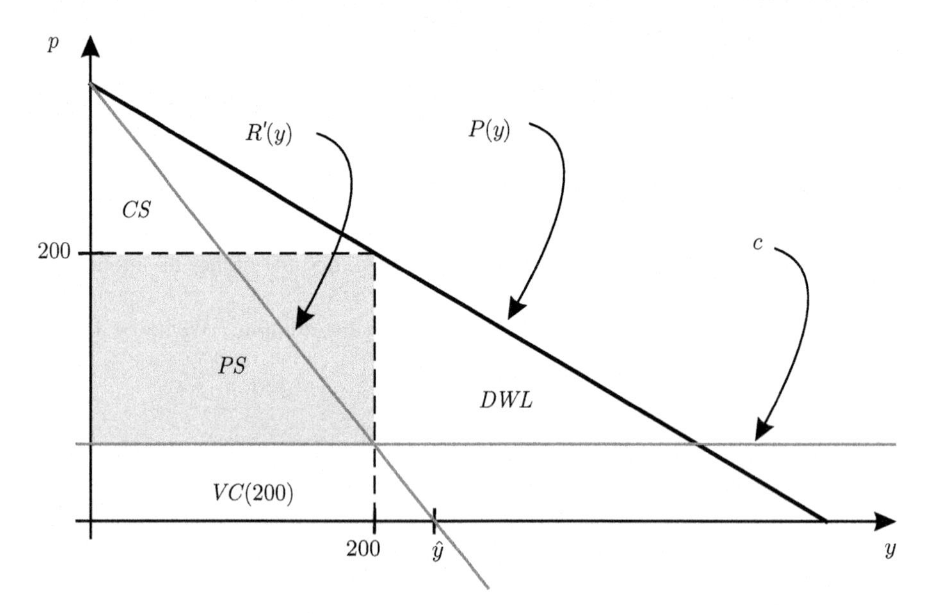

Figure 10.3 Exercise 1.2a). $0 < c < 300$

c) The consumer surplus is

$$CS(y^*) = (300 - 200) \cdot 200 \cdot \frac{1}{2} = 10{,}000,$$

while the producer surplus is

$$PS(y^*) = \pi^* + TFC = 20{,}000.$$

The sum of consumer and producer surplus is thus

$$CS(y^*) + PS(y^*) = 30{,}000.$$

In order to determine the deadweight loss, we must first find the Pareto-efficient quantity $\bar{y}$, which is determined by the intersection of the marginal costs with the (inverse) demand function, i.e.

$$\begin{aligned} MC(y) &= P(y) \\ \Leftrightarrow 100 &= 300 - \frac{1}{2}y \\ \Leftrightarrow \bar{y} &= 400. \end{aligned}$$

Deadweight losses thus amount to

$$\begin{aligned} DWL(y^*) &= \frac{1}{2} \cdot (p^* - MC(y^*)) \cdot (\bar{y} - y^*) \\ &= \frac{1}{2} \cdot (200 - 100) \cdot (400 - 200) = 10{,}000. \end{aligned}$$

10.2.2.2 Solutions to Exercise 2

1. A certain type's willingness to pay $W_j(x)$ is given by the integral over his marginal willingness to pay for all possible values of x,

$$W_j(x) = \int_0^x P_j(z)\,dz, \text{ with } j \in \{L, H\}.$$

It follows that

$$W_L(x) = \begin{cases} 5x - \frac{1}{2}x^2 & \text{for } 0 \leq x \leq 5, \\ 12.5 & \text{else.} \end{cases} \tag{10.4}$$

$$W_H(x) = \begin{cases} 8x - \frac{1}{2}x^2 & \text{for } 0 \leq x \leq 8, \\ 32 & \text{else.} \end{cases}$$

2. Not only does the monopolist know every type's willingness to pay, but he can also identify every individual as being either type L or type H. It is profit maximizing to offer every type j a quantity x_j, such that type j's willingness to pay $W_j(x_j)$ is maximized, and to subsequently set the price p_j for quantity x_j equal to $W_j(x_j)$. The profit of the monopolist is defined as:

$$\pi = p_L + p_H.$$

An individual of type j accepts the contract if his consumer surplus CS_j is (weakly) positive, that is, if

$$CS_L(x_L) \geq 0 \Leftrightarrow W_L(x_L) - p_L \geq 0, \tag{10.5}$$

$$CS_H(x_H) \geq 0 \Leftrightarrow W_H(x_H) - p_H \geq 0. \tag{10.6}$$

Since the monopolist's profit increases in p_j, it follows that

$$W_L(x_L) = p_L, \text{ and} \tag{10.7}$$

$$W_H(x_H) = p_H, \tag{10.8}$$

at the profit maximum. Therefore, the monopolist's maximum profit can be expressed as:

$$\pi(x_L, x_H) = W_L(x_L) + W_H(x_H). \tag{10.9}$$

Taking the partial derivative of Eq. 10.9 with respect to x_L and x_H yields the first-order conditions for a profit maximum:

$$\frac{\partial \pi(x_L, x_H)}{\partial x_L} = 0 \quad \Leftrightarrow \quad 5 - x_L = 0 \quad \Leftrightarrow \quad x_L^* = 5, \tag{10.10}$$

$$\frac{\partial \pi(x_L, x_H)}{\partial x_H} = 0 \quad \Leftrightarrow \quad 8 - x_H = 0 \quad \Leftrightarrow \quad x_H^* = 8. \tag{10.11}$$

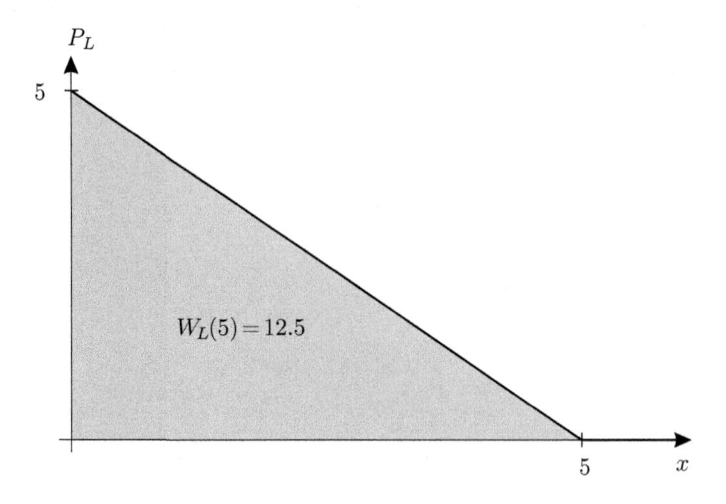

Figure 10.4 Exercise 2.2. $W_L(5)$

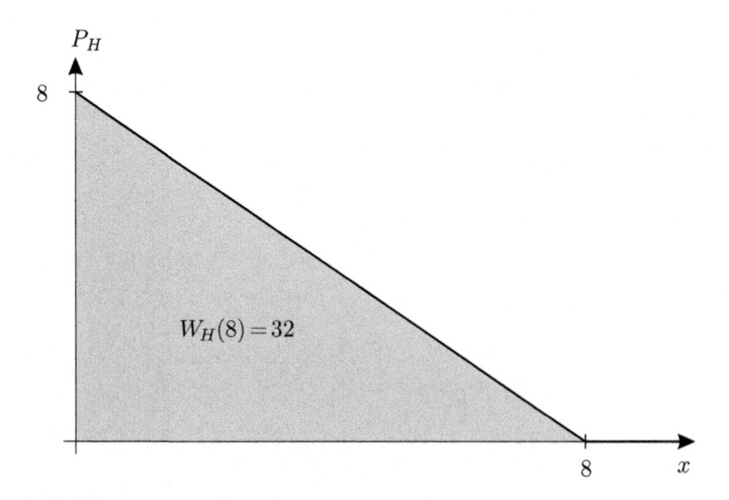

Figure 10.5 Exercise 2.2. $W_H(8)$

From Eq. 10.7 and Eq. 10.8 as well as Eq. 10.10 and Eq. 10.11, it follows that

$$p_L^* = W_L(x_L^*) = 12.5,$$
$$p_H^* = W_H(x_H^*) = 32,$$

as illustrated in Figs. 10.4 and 10.5.
Hence, the contracts are $\{x_L^*, p_L^*\} = \{5, 12.5\}$ and $\{x_H^*, p_H^*\} = \{8, 32\}$. The monopolist's profit amounts to $\pi^* = \pi(x_L^*, x_H^*) = 12.5 + 32 = 44.5$ and the consumer surplus amounts to $CS^* = CS_L(x_L^*) + CS_H(x_H^*) = 0$. Obviously, the

allocation is Pareto-efficient because the sum of the profit and consumer surplus cannot be increased by changing the quantities.

3. The monopolist is no longer able to distinguish between an individual of type L and an individual of type H.
   a) There is only one contract $\{x, p\}$ that is designed such that the monopolist's profit is maximized and a customer of type H will accept it. The monopolist's profit is $\pi = p$.
   An individual of type H will only accept the contract if the consumer surplus is (weakly) positive, that is, if

$$CS_H = W_H(x) - p \geq 0.$$

   Hence, $p = W_H(x)$ holds at the profit maximum and therefore

$$\pi = W_H(x). \tag{10.12}$$

   The first derivative of Eq. 10.12 with respect to x yields the first-order condition for a profit maximum. It is equal to Eq. 10.11 and gives the result $x^* = 8$ and thus, $p^* = W_H(x^*) = 32$. This contract is not attractive for an individual of type L because $W_L(x) = 12.5$ for $x \geq 5$: Type L's willingness to pay does not increase any further if $x > 5$ (see Eq. 10.4). Since $p^* > 12.5$, an individual of type L will not buy the offered contract.
   The resulting profit amounts to $\pi^* = p^* = 32$, and the consumer surplus to $CS = 0$.
   b) There is only one contract $\{x, p\}$ that is designed such that the monopolist's profit is maximized and a customer of type L will accept it. Again, the monopolist's profit can be expressed as $\pi = p$. An individual of type L will only accept the contract if the consumer surplus is (weakly) positive, that is, if

$$CS_L = W_L(x) - p \geq 0.$$

   Hence, $p = W_L(x)$ holds at the profit maximum. Moreover, for any $x > 0$, the WTP of type H is always larger than the WTP of type L. As a consequence, type H will always accept such a contract since $p = W_L(x) < W_H(x)$, for $p > 0$. The monopolist's profit is:

$$\pi(x) = 2 \cdot W_L(x). \tag{10.13}$$

   The first derivative of Eq. 10.13 with respect to x yields the first-order conditions for a profit maximum. We get $x^* = 5$ and $p^* = W_L(x^*) = 12.5$. An individual of type L receives no consumer surplus from the contract because $CS_L(x^*) = 0$. An individual of type H, on the other hand, receives:

$$CS_H(x^*) = W_H(x^*) - p^* = \underbrace{8 \cdot 5 - \frac{1}{2} \cdot 5^2}_{= W_H(x^*)} - \underbrace{12.5}_{= p^*} = 15.$$

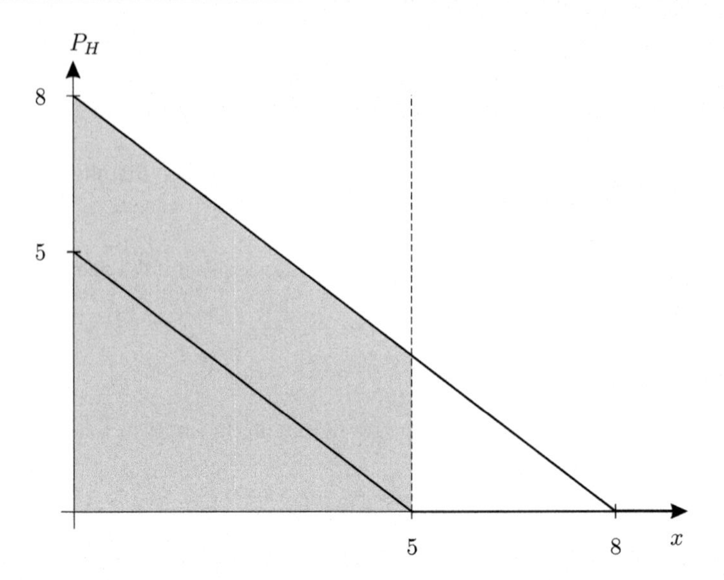

Figure 10.6 Exercise 2.3b). $W_H(5)$

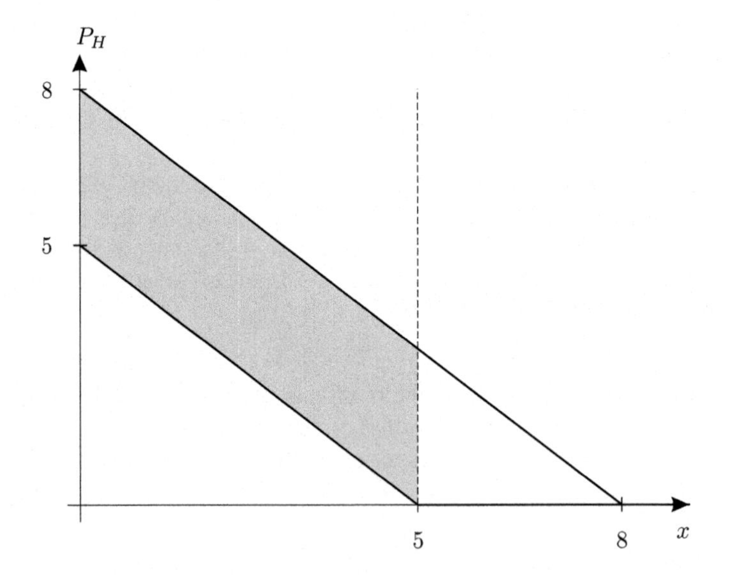

Figure 10.7 Exercise 2.3b). $CS_H(5)$

Fig. 10.6 illustrates the willingness to pay of type H for $x = 5$, while Fig. 10.7 represents the consumer surplus of type H. The total consumer surplus amounts to $CS = CS_L(x^*) + CS_H(x^*) = 0 + 15 = 15$, and the monopolist's profit amounts to $\pi = 2 \cdot p^* = 2 \cdot 12.5 = 25$.

c) There are two contracts with two separate prices. Consequently, the monopolist's profit is

$$\pi = p_L + p_H \tag{10.14}$$

We already calculated the Pareto-efficient quantities in Question 2: $x_L^* = 5$ and $x_H^* = 8$. In order for both types to accept the contracts, the following constraints have to be met:

- the participation constraints, given by Eqs. (10.5) and (10.6);
- the incentive compatibility constraints, i.e. that only an individual of type j will accept the j-contract.

Given these constraints, we can infer the following conditions:

$$W_L(x_L^*) - p_L \geq 0, \tag{10.15}$$
$$W_H(x_H^*) - p_H \geq 0, \tag{10.16}$$
$$W_L(x_L^*) - p_L \geq W_L(x_H^*) - p_H, \tag{10.17}$$
$$W_H(x_H^*) - p_H \geq W_H(x_L^*) - p_L. \tag{10.18}$$

Since the willingness to pay of type L equals 12.5 for $x \geq 5$ (see Eq. 10.4), we know that $W_L(x_H^*) = W_L(x_L^*) = 12.5$. We can utilize this information to modify Eq. 10.17 and we get

$$W_L(x_L^*) - p_L \geq W_L(x_H^*) - p_H \quad \Leftrightarrow \quad p_H \geq p_L.$$

If the price for type H is at least as high as the price for type L, the incentive compatibility constraint for type L is met. For the moment, we assume that this holds true and check again after we arrived at a solution for p_H^* and p_L^* if our solutions do indeed fulfill this condition.

In the next step, we recall that $W_H(x) > W_L(x)$ for all $x > 0$. Utilizing this information in Eq. 10.15, we get

$$W_H(x_L^*) - p_L > 0, \tag{10.19}$$

that a consumer of type H, who accepts an L-contract receives a positive consumer surplus. Inserting Eq. 10.19 into Eq. 10.18 yields $W_H(x_H^*) - p_H > 0$: Accepting an H-contract will always lead to a positive consumer surplus for an individual of type H.

If the quantities for type L and type H as well as the price for the L-contract were already fixed, should condition Eq. 10.18 then hold in its strict form, i.e. $\ldots = \ldots$? Since the monopolist's profit increases in p_H, the price should be chosen in such a way that type H is indifferent between the L-contract and the H-contract. Applying Eq. 10.18, this results in the following price for the H-contract:

$$p_H = W_H(x_H^*) - W_H(x_L^*) + p_L. \tag{10.20}$$

The only missing piece is the price for the L-contract. In Question 3b) we already demonstrated that the monopolist offers the L-contract at a price exactly equal to the willingness to pay of type L. Does this still hold? First

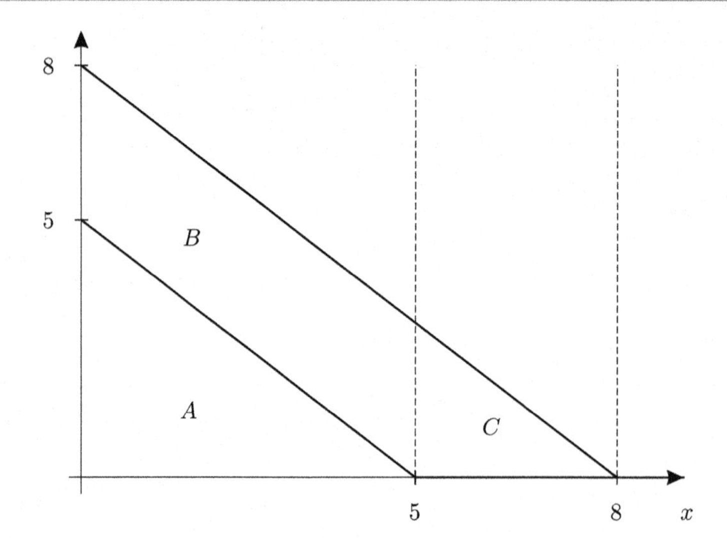

Figure 10.8 Exercise 2.3c)

of all, we take note that any price $p_L > W_L(x_L^*)$ cannot be optimal, because such a contract would not be accepted by an individual of type L. But is there a case such that a price $p_L < W_L(x_L^*)$ is optimal? Compared to $p_L = W_L(x_L^*)$, any L-contract priced at $p_L < W_L(x_L^*)$ would yield a lower profit (see Eq. 10.14). What would be the effect on the H-contract? It follows from Eq. 10.20 that the lower the price p_L, the lower the price p_H. Why? Because a price reduction for the L-contract makes this contract more attractive to a customer of type H. For the incentive compatibility constraints to be fulfilled, the price of the H-contract would have to be lowered as well. But then, the monopolist would get a lower profit from the H-contract. Thus, at the profit maximum, p_L has to be equal to type L's willingness to pay for the Pareto-efficient quantity, that is

$$p_L^* = W_L(x_L^*) = 12.5,$$

which is represented by area A in Fig. 10.8.
This leads to the optimal price for the H-contract (see Eq. 10.20) as

$$p_H^* = W_H(x_H^*) - W_H(x_L^*) + p_L^* = \underbrace{32}_{= W_H(x_H^*)} - \underbrace{27.5}_{= W_H(x_L^*)} + \underbrace{12.5}_{= p_L^*} = 17,$$

which is represented by area $A + C$ in Fig. 10.8. Hence, as assumed, $p_H^* \geq p_L^*$. The profit of the monopolist then becomes

$$\pi^* = \pi(p_L^*, p_H^*) = p_L^* + p_H^* = 12.5 + 17 = 29.5.$$

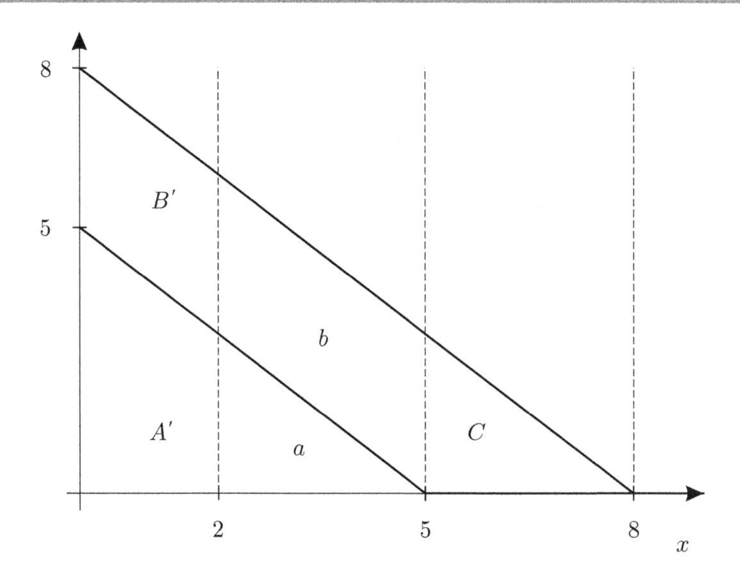

Figure 10.9 Exercise 2.3d)

The consumer surplus corresponds to area B in Fig. 10.8 and is

$$CS = CS_H = W_H(x_H^*) - p_H = 32 - 17 = 15.$$

So far, offering two contracts has not been beneficial to the monopolist, because he was able to generate a profit of $\pi^* = W_H(x^*) = 32$ in Question 3a). Now, taking the contracts $\{x_L^*, p_L^*\} = \{5, 12.5\}$ and $\{x_H^*, p_H^*\} = \{8, 17\}$ as a starting point, let's consider whether a monopolist can increase his profits by tinkering with the quantities offered in the contracts.

d) We reduce the quantity of the L-contract to $\tilde{x}_L = 2$. By conditions (10.15) to (10.18), this will influence the prices p_L and p_H at the profit maximum. We already know from Question 3c) that the price for the L-contract equals the willingness to pay of type L for quantity $\tilde{x}_L$. Therefore,

$$\tilde{p}_L = W_L(\tilde{x}_L) = 5 \cdot 2 - \frac{1}{2} \cdot 2^2 = 8.$$

A reduction in the quantity has to lead to a reduction in p_L, because the participation constraint of type L (see Eq. 10.15) would be violated otherwise. This is illustrated in Fig. 10.9: Induced by the reduction in quantity, the willingness to pay of type L is reduced by an amount represented by area a. Thus, $W_L(\tilde{x}_L) = \tilde{p}_L$, which leads to area A'.

The reduction in price and quantity in the L-contract affects the price of the H-contract. According to Eq. 10.20, prices at the profit maximum are:

$$\tilde{p}_H = W_H(x_H^*) - W_H(\tilde{x}_L) + \tilde{p}_L = \underbrace{32}_{=W_H(x_H^*)} - \underbrace{14}_{=W_H(\tilde{x}_L)} + \underbrace{8}_{=\tilde{p}_L} = 26.$$

What explains the increase in price for the H-contract compared to Question 3c)? Consider Fig. 10.9. Due to the reduction in quantity by $x_L^* - \tilde{x}_L = 3$, the L-contract becomes less attractive to an individual of type H: His willingness to pay is lowered by an amount represented by the area $a + b$. The consumer surplus of type H, if he bought the L-contract, would only be $W_H(\tilde{x}_L) - \tilde{p}_L = 6$ (area B'). This is less than the consumer surplus he receives when buying the H-contract, $W_H(x_H^*) - p_H^* = 15$ (area $B' + b$). The difference between the two corresponds to the loss in consumer surplus for type H, had he bought the L-contract instead of the H-contract. This loss is represented area b in Fig. 10.9.
The monopolist takes advantage of this difference: He increases the price for the H-contract, thus making this contract less attractive. He offers the H-contract at a new price $\tilde{p}_H$, such that type H is again indifferent between the L- and the H-contract. This is to say that the price p_H gets raised by b.
Was this measure beneficial for the monopolist? We compare the reduction in profits induced by the price reduction of the L-contract (area a, $p_L^* - \tilde{p}_L = 4.5$) with the increase in profits due to the price increase in the H-contract (area b, $\tilde{p}_H - p_H^* = 26 - 17 = 9$). The net effect is thus given by $9 - 4.5 = 4.5$. In other words, the monopolist's profit increases by 4.5 compared to Question 3c):

$$\tilde{\pi} = \tilde{p}_L + \tilde{p}_H = 8 + 26 = 34.$$

Thus, offering two contracts has proven beneficial compared to the single contract (see Questions 3a) and 3b)).
The consumer surplus (area B') has been reduced compared to Question 3c):

$$\widetilde{CS} = \widetilde{CS}_H = W_H(x_H^*) - \tilde{p}_H = 32 - 26 = 6.$$

The resulting allocation is no longer Pareto-efficient, as type L no longer gets the Pareto-efficient quantity. This is also reflected by the sum of consumer surplus and profit, which amounts to $34 + 6 = 40$ (compared to 44.5 in the case of a Pareto-efficient allocation).

e) We already know the profit-maximizing level of p_L ($p_L = W_L(x_L^*)$) and the price of the H-contract (see Eq. 10.20) if $x_L = x_L^*$. Thus, for general $x_L \geq 0$, the monopolist's profit becomes:

$$\begin{aligned} \pi = p_L + p_H &= \underbrace{W_L(x_L)}_{=p_L} + \underbrace{W_H(x_H^*) - W_H(x_L) + W_L(x_L)}_{=p_H \text{ according to Eq. 10.20}} \\ &= 2 \cdot W_L(x_L) + W_H(x_H^*) - W_H(x_L). \end{aligned} \tag{10.21}$$

Taking the first derivative of Eq. 10.21 with respect to x_L and setting it equal to zero gives:

$$\pi'(x_L) = 0 \Leftrightarrow 2 \cdot \underbrace{(5 - x_L)}_{=W_L'(x_L)=P_L(x_L)} - \underbrace{(8 - x_L)}_{=W_H'(x_L)=P_H(x_L)} = 0 \Leftrightarrow x_L = 2.$$

We see that we already used the profit-maximizing value of x_L in Question 3d). Hence, profit and consumer surplus are identical to the values calculated in Question 3d).

10.2.2.3 Solutions to Exercise 3

1. We can easily determine the inverse demand function from the given demand function $y_i(p_i)$ (note that $x = y$):

$$x_1(p_1) = 300 - 2\,p_1 \quad \Leftrightarrow \quad P_1(y_1) = 150 - \frac{1}{2}\,y_1,$$
$$x_2(p_2) = 200 - 2\,p_2 \quad \Leftrightarrow \quad P_2(y_2) = 100 - \frac{1}{2}\,y_2.$$

Now we determine the monopolist's profit contingent on y_1 and y_2:

$$\begin{aligned}\pi(y_1, y_2) &= P_1(y_1)\,y_1 + P_2(y_2)\,y_2 - C(y_1 + y_2)\\ &= \left(150 - \frac{1}{2}\,y_1\right)\,y_1 + \left(100 - \frac{1}{2}y_2\right)\,y_2 - c\,y_1 - c\,y_2.\end{aligned} \tag{10.22}$$

At the optimum the marginal revenue is equal to the marginal costs. The partial derivative of the profit function (10.22) with respect to y_1, or y_2, is

$$\frac{\partial\pi(y_1, y_2)}{\partial y_1} = P'(y_1)y_1 + P(y_1) - \frac{\partial C(y_1 + y_2)}{\partial y_1} = 150 - y_1 - c,$$
$$\frac{\partial\pi(y_1, y_2)}{\partial y_2} = P'(y_2)y_2 + P(y_2) - \frac{\partial C(y_1 + y_2)}{\partial y_2} = 100 - y_2 - c.$$

For both markets the optimum is at

$$\frac{\partial\pi(y_1, y_2)}{\partial y_1} = 0 \Leftrightarrow \underbrace{150 - y_1}_{=\text{ marginal revenue}_1} = \underbrace{c}_{=\text{ marginal costs}} \Leftrightarrow y_1^{PD} = 150 - c, \tag{10.23}$$

$$\frac{\partial\pi(y_1, y_2)}{\partial y_2} = 0 \Leftrightarrow \underbrace{100 - y_2}_{=\text{ marginal revenue}_2} = \underbrace{c}_{=\text{ marginal costs}} \Leftrightarrow y_2^{PD} = 100 - c. \tag{10.24}$$

Thus, the price on both markets is

$$p_1^{PD} = P_1(y_1^{PD}) = 150 - \frac{1}{2}\,y_1^{PD} = 150 - \frac{1}{2}\,(150 - c) = 75 + \frac{c}{2},$$
$$p_2^{PD} = P_2(y_2^{PD}) = 100 - \frac{1}{2}\,y_2^{PD} = 100 - \frac{1}{2}\,(100 - c) = 50 + \frac{c}{2}.$$

2. In order for the monopolist to supply both markets, the marginal costs must be smaller than the smaller of both prohibitive prices (the price when $y_i(p_i) = 0$). Comparing the two demand functions, we notice that $c < 100$ has to be fulfilled in order for both markets to be supplied (see also Eq. 10.24).

3. We get the following values for $c = 50$:

$$y_1^{PD} = 150 - c = 100 \qquad \text{and } p_1^{PD} = 75 + \frac{c}{2} = 100,$$

$$y_2^{PD} = 100 - c = 50 \qquad \text{and } p_2^{PD} = 50 + \frac{c}{2} = 75.$$

Thus, the price elasticity of demand at the point p_i^{PD} is (again, note that $x = y$):

$$\varepsilon^{x_1}_{p_1^{PD}} = x_1'(p_1^{PD}) \cdot \frac{p_1^{PD}}{x_1(p_1^{PD})} = -2 \cdot \frac{p_1^{PD}}{x_1^{PD}} = -2 \cdot \frac{100}{100} = -2,$$

$$\varepsilon^{x_2}_{p_2^{PD}} = x_2'(p_2^{PD}) \cdot \frac{p_2^{PD}}{x_2(p_2^{PD})} = -2 \cdot \frac{p_2^{PD}}{x_2^{PD}} = -2 \cdot \frac{75}{50} = -3.$$

As we can tell: The more elastic the demand is at the optimum, the smaller the monopoly price is.

4. Now we assume that price discrimination is prohibited, i.e. the monopolist is forced to offer the good for the identical price on both markets. In this case, we have to aggregate the demand because, from the point of view of the non-price discriminating monopolist, it is a common market. We already know the prohibitive price $\bar{p}_i$ of both markets: $\bar{p}_1 = 150$, while $\bar{p}_2 = 100$. Hence, if $100 \leq p < 150$, then total demand equals the demand in the first market ($X(p) = x_1(p)$), while for $p < 100$ total demand equals the sum over both markets ($X(p) = x_1(p) + x_2(p)$):

$$X(p) = \begin{cases} x_1(p), & \text{for } 100 < p \leq 150, \\ x_1(p) + x_2(p), & \text{for } 0 \leq p \leq 100. \end{cases}$$

$$= \begin{cases} 300 - 2\,p, & \text{for } 100 < p \leq 150, \\ 500 - 4\,p, & \text{for } 0 \leq p \leq 100. \end{cases}$$

However, because we want to maximize the profit of the monopolist via the quantity (as before), we need the corresponding inverse demand function $P(y)$:

$$P(y) = \begin{cases} 150 - \frac{1}{2}y, & \text{for } 0 \leq y < 100, \\ 125 - \frac{1}{4}y, & \text{for } 100 \leq y \leq 500. \end{cases}$$

Then, the monopolist's profit function becomes

$$\pi(y) = P(y)\,y - C(y) = \begin{cases} (150 - \frac{1}{2}\,y)\,y - c\,y, & \text{for } 0 \leq y < 100, \\ (125 - \frac{1}{4}\,y)\,y - c\,y, & \text{for } 100 \leq y \leq 500. \end{cases}$$

The marginal revenue is

$$R'(y) = P'(y)\,y + P(y) = \begin{cases} 150 - y, & \text{for } 0 \leq y < 100, \\ 125 - \frac{1}{2}\,y, & \text{for } 100 \leq y < 500. \end{cases}$$

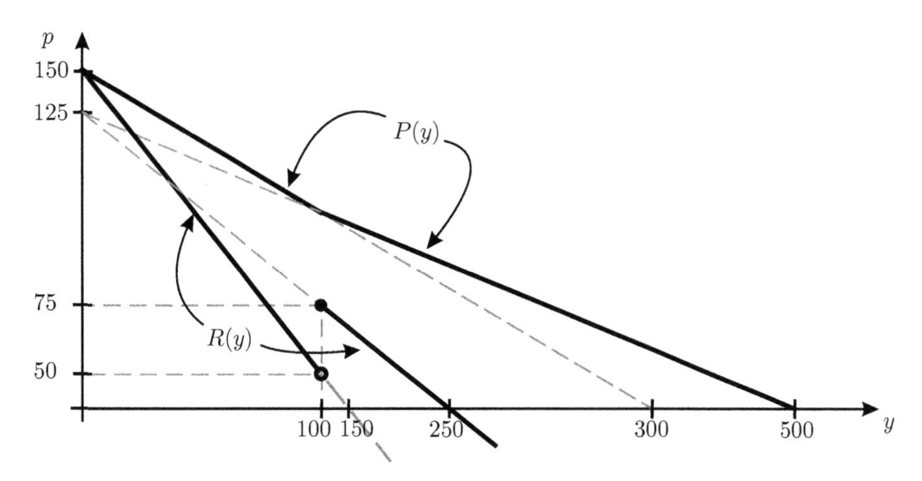

Figure 10.10 Exercise 3.4. Aggregate demand

This is illustrated in Fig. 10.10. We see that, for $y < 100$ the corresponding monopoly price is larger than 100 and thus, only the consumers from the first market will demand the good. The marginal revenue thus corresponds the marginal profit in the first market. If $y > 100$ the monopoly price drops below 100 and the second market is also supplied. Thus, the marginal revenue corresponds to the marginal revenue of the entire market. The discontinuity of the marginal revenue function at $y = 100$ is due to the change in the price elasticity of demand at $y = 100$. The demand immediately becomes more elastic for $y > 100$.

5. For the monopolist the following is valid at the optimum:

$$\pi'(y) = 0 \Leftrightarrow R'(y) = C'(y) \Leftrightarrow \begin{cases} 150 - y = c, & \text{for } 0 \leq y < 100, \\ 125 - \frac{1}{2}\, y = c, & \text{for } 100 \leq y \leq 500. \end{cases} \tag{10.25}$$

Thus, we get (for $c \geq 0$) two results:

(i) For $y < 100$ it follows from (10.25) that $y^* = y_1^{PD} = 150 - c$ and $p^* = p_1^{PD} = 75 + \frac{c}{2}$ for $c > 50$.

(ii) For $y \geq 100$ it follows from (10.25) that $y^* = 250 - 2\,c$ and $p^* = 62.5 + \frac{c}{2}$ for $c \leq 75$.

What do we learn from (i) and (ii)? Apparently, there are three different intervals to be distinguished:

1. **Interval**: $150 > c > 75$. In this case there is a unique optimum at $y^* = 150 - c$. Hence, only the first market will be supplied, since the monopoly price ($p^* = 75 + \frac{c}{2}$) in this interval is larger than 100. This case is represented by Fig. 10.11 for $c = 90$. For this value of c we get $y^* = 60$ and $p^* = 120$.
2. **Interval**: $0 \leq c \leq 50$. In this case there is unique optimum with $y^* = 250 - 2\,c$. Hence, both markets will be supplied, since the monopoly price is

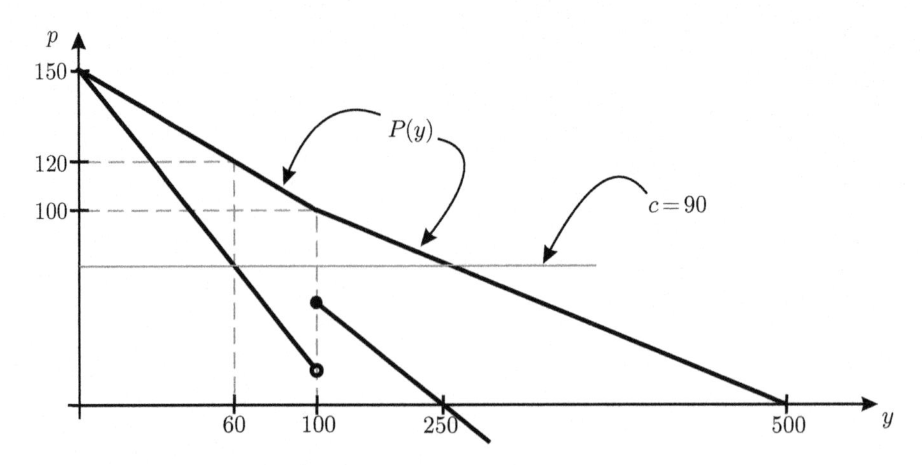

Figure 10.11 Exercise 3.5. Monopoly without PD for $c = 90$

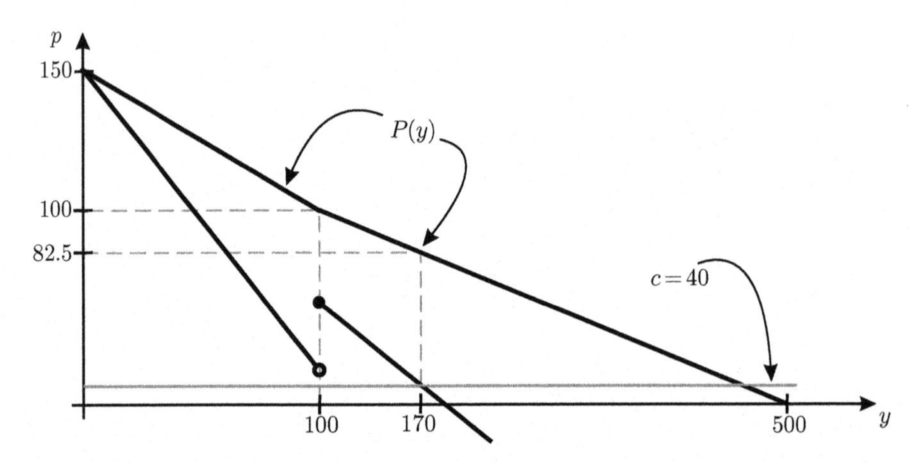

Figure 10.12 Exercise 3.5. Monopoly without PD for $c = 40$

below 100 ($p^* = 62.5 + \frac{c}{2}$) in this case. This case is represented by Fig. 10.12 for $c = 40$. For this value of c we get $y^* = 170$ and $p^* = 82.5$.

3. **Interval**: $50 < c \leq 75$. In this case there are two (local) optima, since the first order condition for a profit maximum ($R'(y) = C'(y)$) is fulfilled for two distinct values of y: At $y = 150 - c$ and at $y = 250 - 2c$ (see Eq. 10.25). This case is represented by Fig. 10.13 for $c = 65$. For this value of c we get as the first optimum $y^* = 85$ and $p^* = 107.5$, and as the second optimum $y^* = 120$ and $p^* = 95$.
   At $y = 85$ there is one optimum, where only the first market will be supplied, at $y = 120$ there is an optimum where both market will be supplied. Which

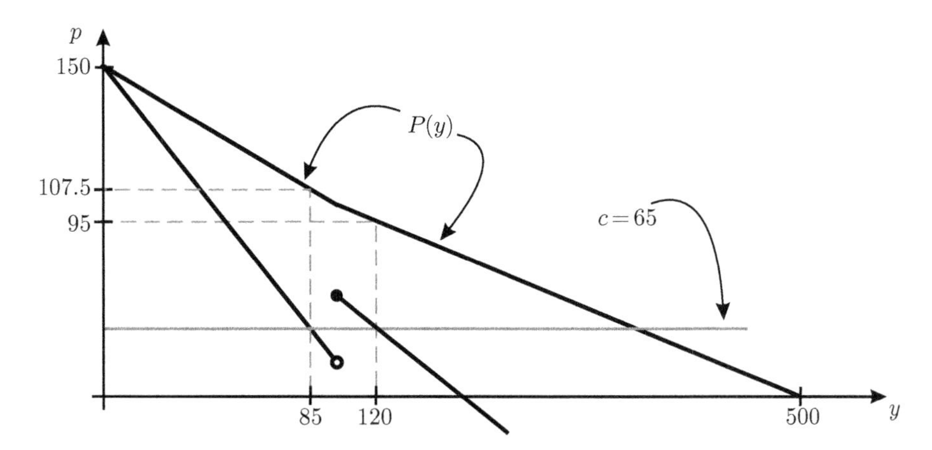

Figure 10.13 Exercise 3.5. Monopoly without PD for $c = 65$

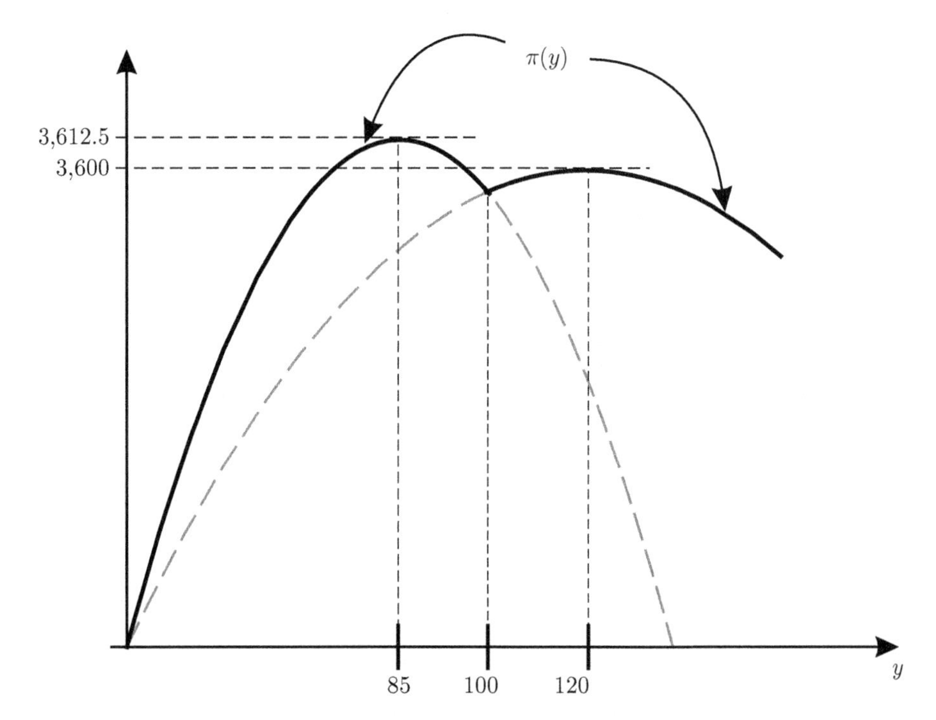

Figure 10.14 Exercise 3.5. Monopoly without PD for $c = 65$

value of y does now maximize the monopolist's profit? To answer this question, we need to compare the monopolist's profit in both cases. Figure 10.14 illustrates the monopolist's profit contingent on the value of y. Apparently,

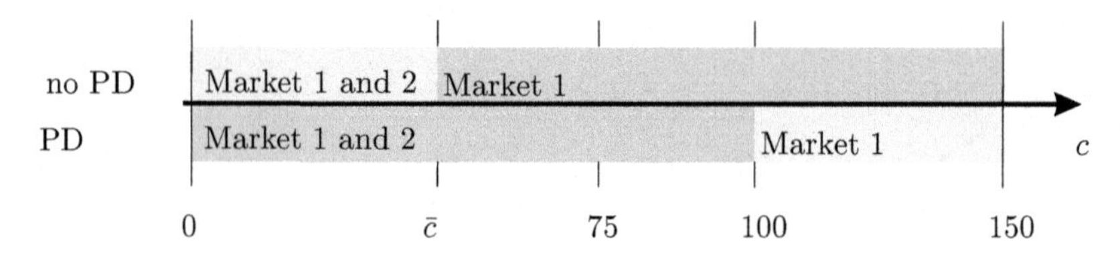

Figure 10.15 Exercise 3.5. Comparison

the profit at $y = 85$ exceeds the profit at $y = 120$:

$$\begin{aligned} \pi(85) &= \underbrace{85 \cdot 107.5}_{=\text{revenue}} - \underbrace{85 \cdot 65}_{=\text{cost}} = 3{,}612.5 \\ > \pi(120) &= \underbrace{120 \cdot 95}_{=\text{revenue}} - \underbrace{120 \cdot 65}_{=\text{cost}} = 3{,}600. \end{aligned} \tag{10.26}$$

Thus, the monopolist would only supply the first market at the profit maximum.
Next, we want to analyze, whether this finding holds for all values of c between 50 and 75: The following holds if the monopolist is indifferent between both scenarios:

$$\begin{aligned} \pi(150 - c) &= \pi(250 - 2\,c) \\ (150 - c)\left(75 + \frac{c}{2}\right) - c \cdot (150 - c) &= (250 - 2\,c)\left(62.5 + \frac{c}{2}\right) - c \cdot (250 - 2\,c) \\ (150 - c)\left(75 - \frac{c}{2}\right) &= (250 - 2\,c)\left(62.5 - \frac{c}{2}\right) \\ 8{,}750 - 200\,c + c^2 &= 0. \end{aligned} \tag{10.27}$$

This quadratic equation can be solved by means of the quadratic solution formula. We then get the critical value of c, $\bar{c} \approx 64.64$. Since we already know that the profit is larger at $c = 65$ if the monopolist serves only the first market (see Eq. 10.26), we conclude that for $c > \bar{c}$ only market 1 will be served, whereas for $c \leq \bar{c}$ both markets will be served.[1]

To sum up, we learned that the monopolist in case price discrimination is banned, only serves both markets if c is sufficiently small, i.e. $c \leq \bar{c}$. In case price discrimination is not banned, however, he will supply both markets unless marginal costs are larger than the prohibitive price in the second market, i.e. $c \geq 100$. This relation is illustrated in Fig. 10.15.

[1] Actually, the monopolist is indifferent at $c = \bar{c}$ between serving both and serving only one market. We assume that in this case he will serve both markets.

10.3 Multiple Choice

10.3.1 Problems

10.3.1.1 Exercise 1

A monopolist is confronted with an inverse market demand function $P(y) = 101 - y$. He produces without any fixed costs and with marginal costs of $MC(y) = c$.

1. Determine the profit-maximizing quantity.
   a) $y^* = \frac{101-c}{2}$.
   b) $y^* = 101 - c$.
   c) $y^* = \frac{100-c}{2}$.
   d) $y^* = 100 - c$.
   e) None of the above answers are correct.
2. For which values of $\tilde{y}$ would the monopolist's marginal revenues be negative?
   a) $\tilde{y} > 50.5$.
   b) $\tilde{y} > 50$.
   c) $\tilde{y} > 101$.
   d) The marginal revenues cannot be negative.
   e) None of the above answers are correct.
3. Determine the profit-maximizing quantity and the profit if the monopolist can utilize perfect price discrimination.
   a) $y^* = 101 - c$, the profit is indeterminate because the price is indeterminate.
   b) $y^* = \frac{101-c}{2}$ and $\pi(y^*) = \frac{101-c}{2}$.
   c) $y^* = 101 - c$ and $\pi(y^*) = 101 - c$.
   d) $y^* = 101 - c$ and $\pi(y^*) = \frac{(101-c)^2}{2}$.
   e) None of the above answers are correct.
4. Determine the consumer surplus with and without price discrimination based on Questions 1 and 3.
   a) With PD: $CS(y^*) = \frac{(101-c)^2}{2}$. Without PD: $CS(y^*) = 0$.
   b) With PD: $CS(y^*) = 0$. Without PD: $CS(y^*) = \frac{(101-c)^2}{8}$.
   c) With PD: $CS(y^*) = 0$. Without PD: $CS(y^*) = \frac{(101-c)^2}{4}$.
   d) With PD: $CS(y^*) = \frac{(101-c)^2}{2}$. Without PD: $CS(y^*) = \frac{(101-c)^2}{4}$.
   e) None of the above answers are correct.

10.3.1.2 Exercise 2

A monopolist is confronted with the inverse demand function $P(y) = 11 - y$. The cost function is $C(y) = y$.

1. Calculate the profit-maximizing supply as well as the profit if price discrimination is not possible.
   a) $y^* = 5$ and $\pi(y^*) = 25$.
   b) $y^* = 6$ and $\pi(y^*) = 25$.
   c) $y^* = 5$ and $\pi(y^*) = 30$.

d) $y^* = 6$ and $\pi(y^*) = 30$.
e) None of the above answers are correct.

The monopolist has the option of investing in advertising in order to influence the demand of the product. Investments of F lead to a new demand of $P(y) = 13 - y$.

2. Determine the profit-maximizing quantity and profit, without taking into account the investment costs F, if price discrimination is not possible.
   a) $y^* = 8$ and $\pi(y^*) = 32$.
   b) $y^* = 6$ and $\pi(y^*) = 36$.
   c) $y^* = 8$ and $\pi(y^*) = 36$.
   d) $y^* = 8$ and $\pi(y^*) = 50$.
   e) None of the above answers are correct.
3. When would the monopolist carry out this investment?
   a) Never.
   b) Only if $F \leq 7$.
   c) Only if $F = 10$.
   d) Only if $F \leq 6$.
   e) None of the above answers are correct.
4. When would the monopolist carry out this investment if she could perfectly discriminate prices?
   a) Never.
   b) Only if $F = 22$.
   c) Only if $F \leq 72$.
   d) Only if $F \leq 22$.
   e) None of the above answers are correct.

10.3.1.3 Exercise 3

A monopolist has the cost function $C(y) = 2\,y$ and is confronted with the inverse demand function $P(y) = 200 - y$.

1. Determine the monopolist's optimal quantity and optimal price if price discrimination is not possible.
   a) The optimal quantity is $y^* = 99$, the optimal price is $p^* = 101$.
   b) The optimal quantity is $y^* = 85$, the optimal price is $p^* = 115$.
   c) The optimal quantity is $y^* = 101$, the optimal price is $p^* = 99$.
   d) The optimal quantity is $y^* = 111$, the optimal price is $p^* = 89$.
   e) None of the above answers are correct.
2. Determine the profit and the consumer surplus without price discrimination.
   a) Profits are 9,200 and consumer surplus is 4,600.
   b) Profits are 9,800 and consumer surplus is 5,000.
   c) Profits are 9,801 and consumer surplus is 4,600.
   d) Profits are 9,801 and consumer surplus is $\frac{9{,}801}{2}$.
   e) None of the above answers are correct.

3. Determine the monopolist's optimal quantity and price with perfect price discrimination.
   a) The optimal quantity is $y^{*PD} = 200$, the optimal price is $p^{*PD} = 2$.
   b) The optimal quantity is $y^{*PD} = 198$, the optimal price corresponds to the individual consumer's willingness to pay.
   c) The optimal quantity is $y^{*PD} = 98$, the optimal price corresponds to the individual consumer's willingness to pay if it is larger or equal to 2.
   d) The optimal quantity is $y^{*PD} = 198$, the optimal price is $p^{*PD} = 2$.
   e) None of the above answers are correct.
4. Determine profits and consumer surplus with perfect price discrimination.
   a) Profits are 39,204 and consumer surplus is 796.
   b) Profits are 39,601 and consumer surplus is 0.
   c) Profits are 39,204 and consumer surplus is 0.
   d) Profits are 0 and consumer surplus is 39,601.
   e) None of the above answers are correct.

Due to a free-trade agreement with a neighboring country, the monopolist can supply a second market. That market has demand $P(\tilde{y}) = 100 - \tilde{y}$, where $\tilde{p}$ is the price in that market and $\tilde{y}$ is the quantity of the good traded in that market.

5. Determine the monopolist's optimal quantity and price in both markets if third-degree price discrimination is possible.
   a) In market 1, the optimal quantity is $y^* = 99$ and the optimal price is $p^* = 101$. In market 2, the optimal quantity is $\tilde{y}^* = 49$ and the optimal price is $\tilde{p}^* = 51$.
   b) In market 1, the optimal quantity is $y^* = 85$ and the optimal price is $p^* = 115$. In market 2, the optimal quantity is $\tilde{y}^* = 51$ and the optimal price is $\tilde{p}^* = 49$.
   c) In market 1, the optimal quantity is $y^* = 101$ and the optimal price is $p^* = 99$. In market 2, the optimal quantity is $\tilde{y}^* = 60$ and the optimal price is $\tilde{p}^* = 40$.
   d) In market 1, the optimal quantity is $y^* = 111$ and the optimal price is $p^* = 89$. In market 2, the optimal quantity is $\tilde{y}^* = 30$ and the optimal price is $\tilde{p}^* = 70$.
   e) None of the above answers are correct.

10.3.1.4 Exercise 4

A monopolist with cost function $C(y) = 2(y_1 + y_2)$ supplies two different markets, 1 and 2, where y_1 is the quantity supplied to one and y_2 is the quantity supplied to the other market. The inverse demand functions for the two markets are $P_1(y_1) = 84 - y_1$ and $P_2(y_2) = 42 - y_2$.

1. Determine the monopolist's optimal quantities and prices with third-degree price discrimination.
   a) The optimal quantities are $y_1^* = 48$ and $y_2^* = 20$, the prices are $p_1^* = 34$ and $p_2^* = 22$.
   b) The optimal quantities are $y_1^* = 41$ and $y_2^* = 20$, the prices are $p_1^* = 43$ and $p_2^* = 22$.
   c) The optimal quantities are $y_1^* = 41$ and $y_2^* = 25$, the prices are $p_1^* = 41$ and $p_2^* = 22$.
   d) The optimal quantities are $y_1^* = 31$ and $y_2^* = 18$, the prices correspond to the individual consumer's willingness to pay in markets 1 and 2.
   e) None of the above answers are correct.
2. A regulation authority forbids price discrimination. Determine the optimal quantity and the optimal price for the new situation.
   a) The optimal quantity is $y^* = 61$, the optimal price is $p^* = 43$.
   b) The optimal quantity is $y^* = 50$, the optimal price is $p^* = 20$.
   c) The optimal quantity is $y^* = 55$, the optimal price is $p^* = 32.5$.
   d) The optimal quantity is $y^* = 61$, the optimal price is $p^* = 32.5$.
   e) None of the above answers are correct.
3. Now, assume that the monopolist can discriminate prices perfectly. Determine the optimal quantity and the price function.
   a) The optimal quantity is $y^* = 80$, every consumer with a willingness to pay above 2 pays the amount she is willing to pay, while all others do not consume the good.
   b) The optimal quantity is $y^* = 110$, every consumer with a willingness to pay above 1 pays the amount she is willing to pay, while all others do not consume the good.
   c) The optimal quantity is $y^* = 122$, every consumer with a willingness to pay above 2 pays pays the amount she is willing to pay, while all others do not consume the good.
   d) The optimal quantity is $y^* = 98$, the optimal price is $p^* = 20$.
   e) None of the above answers are correct.
4. Determine the sum of producer and consumer surplus from Question 3.
   a) The sum of consumer and producer surplus is 5,250.
   b) The sum of consumer and producer surplus is 3,625.
   c) The sum of consumer and producer surplus is 4,162.
   d) The sum of consumer and producer surplus is 4,872.
   e) None of the above answers are correct.

10.3.1.5 Exercise 5

A profit-maximizing monopolist faces two types of buyers: H and L. The demand functions of both types are shown in Figs. 10.16 and 10.17. Both types occur in the population with equal probability. The monopolist can produce with zero marginal costs, while fixed costs are low enough to maintain production in all of the following situations.

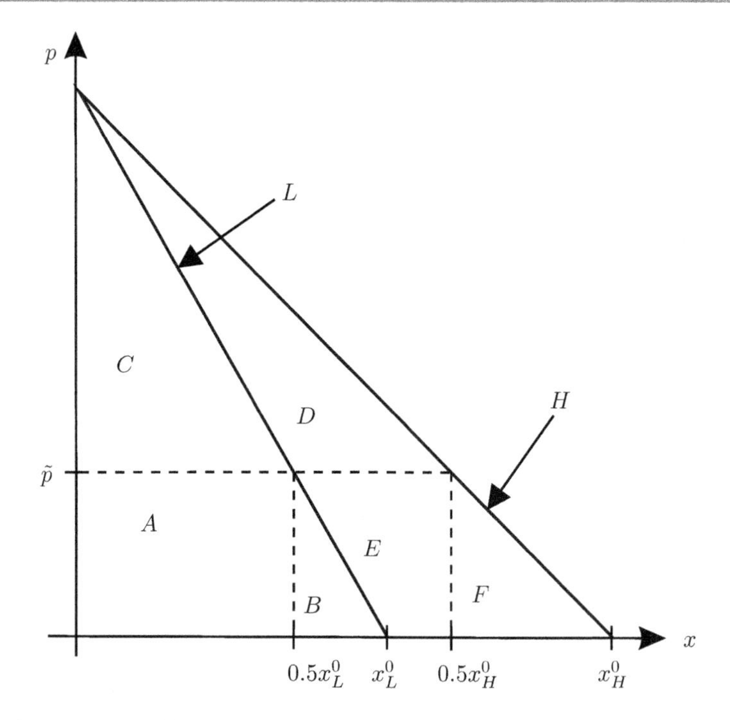

Figure 10.16 Exercise 5.2

1. Consider Fig. 10.16: Determine the optimal quantity sold by the monopolist to both types (x_L^* and x_H^*) if the monopolist has perfect information about the type of each buyer. Which are the associated prices (p_L and p_H)?
   a) The optimal quantities are $x_L^* = 0.5x_L^0$ and $x_H^* = 0.5x_H^0$, the corresponding prices are $p_L = A + C$ and $p_H = A + B + C + D + E$.
   b) The optimal quantities are $x_L^* = x_L^0$ and $x_H^* = x_H^0$, the corresponding prices are $p_L = A + B + C$ and $p_H = A + B + C + D + E + F$.
   c) The optimal quantities are $x_L^* = x_L^0$ and $x_H^* = x_H^0$, the corresponding prices are $p_L = A + B + C$ and $p_H = D + E + F$.
   d) The optimal quantities are $x_L^* = 0.5x_L^0$ and $x_H^* = 0.5x_H^0$, the corresponding prices are $p_L = A + C$ and $p_H = D + E$.
   e) None of the above answers are correct.
2. Consider Fig. 10.16: What is the average profit in a case of perfect information?
   a) The average profit is $A + B + C + 0.5\,(D + E + F)$.
   b) The average profit is $A + C + 0.5\,(A + B + C + D + E)$.
   c) The average profit is $A + B + C + D + E + F$.
   d) The average profit is $A + C + D + E$.
   e) None of the above answers are correct.

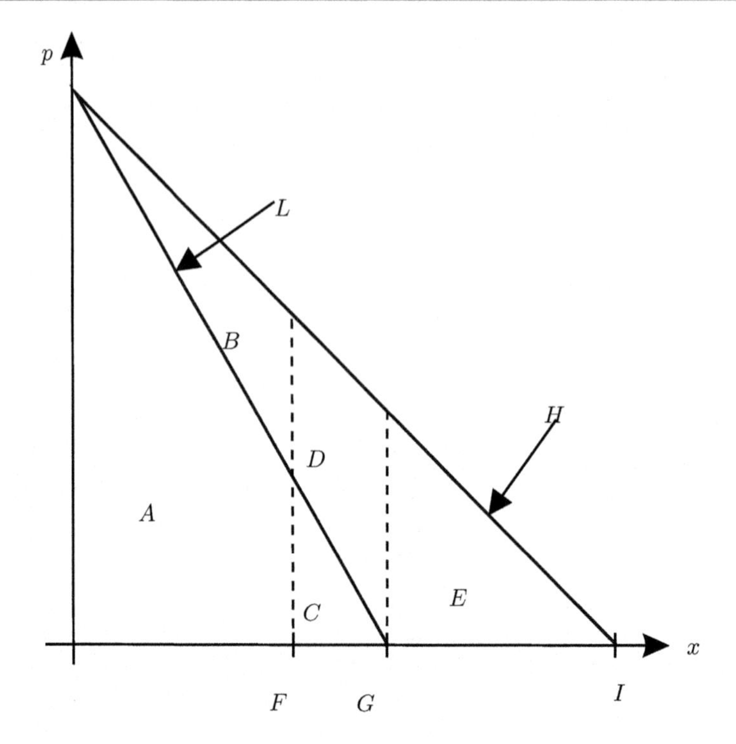

Figure 10.17 Exercise 5.4

3. Consider Fig. 10.17: Assume the monopolist does not know the type of consumer. Determine the optimal quantity the monopolist would sell to the customer of type L at the profit-maximizing output level (x_L^*).
   a) $x_L^* = G$.
   b) $x_L^* = F$.
   c) $x_L^* = I$.
   d) $x_L^* = A + C$.
   e) None of the above answers are correct.
4. Consider Fig. 10.17: Assume further that the monopolist does not know the type of consumer. At which price would the monopolist sell to the consumer of type H at the profit-maximizing output level?
   a) The price is $A + C + B + D + E$.
   b) The price is $B + D + E$.
   c) The price is $A + C + B + D$.
   d) The price is $A + C + D + E$.
   e) None of the above answers are correct.

10.3.2 Solutions

10.3.2.1 Solutions to Exercise 1

- Question 1, answer a) is correct.
- Question 2, answer a) is correct.
- Question 3, answer d) is correct.
- Question 4, answer b) is correct.

10.3.2.2 Solutions to Exercise 2

- Question 1, answer a) is correct.
- Question 2, answer b) is correct.
- Question 3, answer e) is correct. The correct answer is only if $F \leq 11$.
- Question 4, answer d) is correct.

10.3.2.3 Solutions to Exercise 3

- Question 1, answer a) is correct.
- Question 2, answer d) is correct.
- Question 3, answer b) is correct.
- Question 4, answer e) correct. The correct answer is a profit of 19,602 and a consumer surplus of 0.
- Question 5, answer a) is correct.

10.3.2.4 Solutions to Exercise 4

- Question 1, answer b) is correct.
- Question 2, answer d) is correct.
- Question 3, answer c) is correct.
- Question 4, answer c) is correct.

10.3.2.5 Solutions to Exercise 5

- Question 1, answer b) is correct.
- Question 2, answer a) is correct.
- Question 3, answer b) is correct.
- Question 4, answer d) is correct.

Principles of Game Theory 11

11.1 True or False

11.1.1 Statements

11.1.1.1 Block 1

Consider the following sequential game (Fig. 11.1). Player 1 has the strategies {No entry, Entry}, while player 2 has the strategies {Fight, Concede}.

1. (No entry, Fight) is a Nash equilibrium.
2. (No entry, Concede) is a Nash equilibrium.
3. (Entry, Fight) is a Nash equilibrium.
4. (Entry, Concede) is a Nash equilibrium.

11.1.1.2 Block 2

Consider the following game in normal form (Table 11.1).

1. Strategy U is dominant for player 1.
2. (D, R) is a Nash equilibrium in this game.
3. (U, L) is a Nash equilibrium in this game.
4. (D, R) is an equilibrium in dominant strategies in this game.

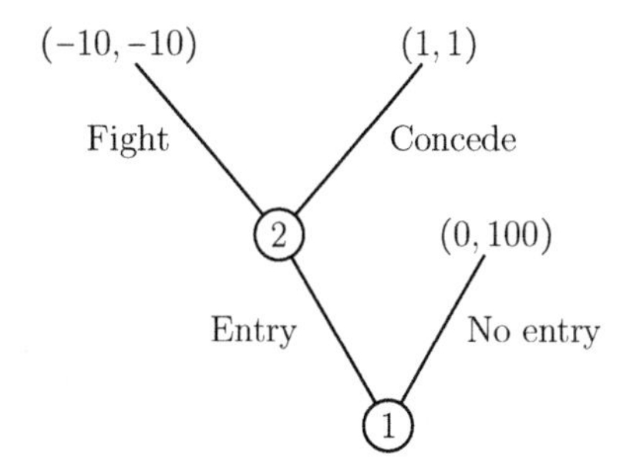

Figure 11.1 Block 1. A game tree

M. Kolmar, M. Hoffmann, *Workbook for Principles of Microeconomics*,
Springer Texts in Business and Economics, https://doi.org/10.1007/978-3-319-62662-8_11

Table 11.1 Block 2. A 2 × 2 Matrix

		Player 2	
		L	R
Player 1	U	$-1, -1$	$15, 0$
	D	$0, 15$	$-10, -10$

Table 11.2 Block 3, Statement 4. A game in normal form

		Player 2			
		OO	OU	UO	UU
Player 1	Y	$1, 0$	$1, 0$	$1, 1$	$1, 1$
	X	$1, 1$	$1, 2$	$1, 1$	$1, 2$

11.1.1.3 Block 3

Consider the following game in extensive form (Fig. 11.2).

1. The strategy sets of the players are $S_1 = \{Y, X\}$ for player 1 and $S_2 = \{O, U\}$ for player 2.
2. In order to maximize his utility, player 2 will never choose O.
3. This is a simultaneous-move game.
4. The following game in normal form (Table 11.2) has the same Nash equilibrium/equilibria as the former extensive-form game.

11.1.1.4 Block 4

1. A game in strategic form can be completely described by the number of players, the strategy sets, and a utility function for each player and each possible outcome.
2. In the so-called "ultimatum game," player A can offer player B any amount of money between 0 and 10 Swiss Francs. Player B can accept or decline. If she declines, both players receive 0; if she accepts, the players receive the allocation suggested by A. Assume that the utility of the players corresponds to the amount of money they receive. In that case, it is a Nash equilibrium if player A offers zero and player B accepts every offer.
3. A game in normal form always has at least one Nash equilibrium in pure strategies.
4. The sequential-move game illustrated in Fig. 11.3 always has at least one Nash equilibrium.

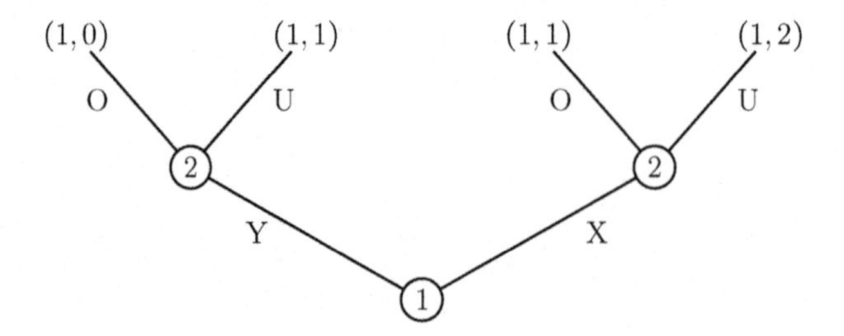

Figure 11.2 Block 3. A game tree

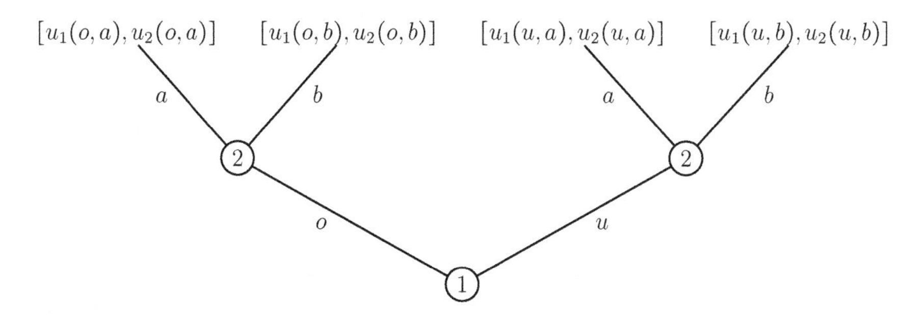

Figure 11.3 Block 4, Statement 4. A game tree

Table 11.3 Block 6, Statements 3 and 4. A normal-form game

		Driver 2	
		L	R
Driver 1	L	10, 10	−10, −10
	R	−10, −10	10, 10

11.1.1.5 Block 5

Consider the prisoner's dilemma (see Table 11.7 in Chapter 11.4.2).

1. The Pareto-efficient strategy combination is an equilibrium if the players can communicate with each other and agree on a cooperation strategy.
2. In this game, there is exactly one strategy profile that is not Pareto-efficient.

Consider a coordination game like "Meeting in New York" (see Table 11.5 in Chapter 11.4).

3. No predictions concerning the outcome of this game can be made.
4. The game "Meeting in New York" has no equilibrium in dominant strategies.

11.1.1.6 Block 6

1. If for all players a reaction function exists, a Nash equilibrium in pure strategies exists.
2. Assume that two players, 1 and 2, have two strategies, a and b each. Every strategy combination yields the same payoffs. Then, no Nash equilibrium exists.

Cars can drive on the right side R or the left side L of the street. The situation between two drivers, 1 and 2, can be illustrated by the following game in normal form (Table 11.3).

3. There is an equilibrium in dominant strategies.
4. There are exactly two Nash equilibria in pure strategies.

Table 11.4 Block 7. A normal-form game

		Player 2	
		L	*R*
Player 1	*U*	100, 100	50, 0
	D	100, 100	0, 150

Table 11.5 Block 8. A normal-form game

		Player 2	
		L	*R*
Player 1	*U*	a_1, a_2	c_1, c_2
	D	b_1, b_2	d_1, d_2

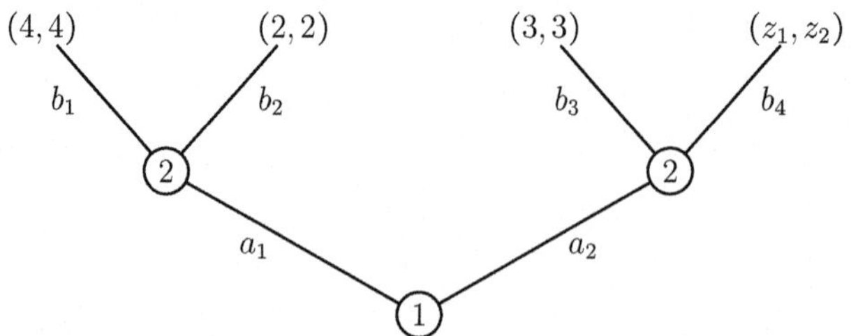

Figure 11.4 Block 9. A game tree

11.1.1.7 Block 7

Examine the following game in normal form (Table 11.4).

1. Strategy D is dominant for player 1.
2. (D, L) is a Nash equilibrium in this game.
3. (U, L) is a Nash equilibrium in this game.
4. (D, R) is a Nash equilibrium in this game.

11.1.1.8 Block 8

Consider the following normal-form game (Table 11.5).

1. There is always a Nash equilibrium in pure strategies, independently of the parameters of the game $a_1, a_2, \ldots, d_2$.
2. Assume that $a_1 = a_2 = 1$, $b_1 = b_2 = 2$, $c_1 = c_2 = 3$ and $d_1 = d_2 = 4$. (D, R) is an equilibrium in dominant strategies.
3. Assume that $a_1 = a_2 = 4$, $b_1 = b_2 = 2$, $c_1 = c_2 = 3$ and $d_1 = d_2 = 4$. The game has exactly one Nash equilibrium in pure strategies.
4. Assume that $a_1 = a_2 = 1$, $b_1 = b_2 = 2$, $c_1 = c_2 = 3$ and $d_1 = d_2 = 1$. Both (D, L) and (U, R) are Nash equilibria in this game.

11.1.1.9 Block 9

Consider the following game tree (Fig. 11.4).

1. The strategy sets are $\{a_1, a_2\}$ for player 1 and $\{b_1, b_2, b_3, b_4\}$ for player 2.
2. Independent of the parameters of the game, z_1, z_2, there is always a Nash equilibrium in pure strategies.

3. Assume that $z_1 = z_2 = 4$. The game has four Nash equilibria in pure strategies.
4. Assume that $z_1 = z_2 = 1$. There is one Nash equilibrium in the game.

11.1.1.10 Block 10

1. All Pareto-dominant equilibria in an extensive-form game can be found by means of the concept of backward-induction.
2. If, in a game with two players, a strategy A is the best response to another strategy B, and this strategy B is also the best response to the former strategy A, then (A, B) is a Nash equilibrium.
3. A strategy profile that Pareto-dominates all other strategy profiles is a Nash equilibrium.
4. In coordination games, focal points can help one coordinate on a strategy profile that is a Nash equilibrium.

11.1.1.11 Block 11

1. A Nash equilibrium is always an equilibrium in dominant strategies.
2. An equilibrium in dominant strategies is always a Nash equilibrium.
3. A dominant strategy is always a best response to all possible strategy profiles of all other players.
4. In extensive-form games, Nash equilibria can be based on non-credible threats.

11.1.1.12 Block 12

1. A game consists of a specification of the players, their strategies, and a mapping from strategies onto utilities.
2. A game consists of a specification of the players, their strategies, a mapping from strategies onto outcomes, and a mapping from outcomes onto utilities.
3. The reaction function defines the strategies that maximize a player's utility for one of the other players' strategy profiles.
4. If there exists a reaction function for each player, then there also exists an equilibrium in dominant strategies.

11.1.1.13 Block 13

1. The "dilemma" that prisoners in the prisoner's dilemma face is based on the inability to anticipate the strategy the other player will choose because there are multiple Nash equilibria in the game.
2. Coordination games have multiple Nash equilibria.
3. If a game has an equilibrium in dominant strategies, this equilibrium must be Pareto-efficient, because every player chooses the strategy with the highest utility.
4. The market-entry game in extensive form can also be portrayed in normal form.

Table 11.6 Block 14. A game in normal form

		Player 2	
		C	*D*
Player 1	*A*	5,4	3,3
	B	4,4	4,5

Table 11.7 Block 1. A 2×2 Matrix

		Player 2	
		Fight	Concede
Player 1	Entry	$-10, -10$	$\underline{1}, \underline{1}$
	No entry	$\underline{0}, \underline{100}$	$0, \underline{100}$

Table 11.8 Block 2. A 2×2 Matrix

		Player 2	
		L	*R*
Player 1	*U*	$-1, -1$	$\underline{15}, \underline{0}$
	D	$\underline{0}, \underline{15}$	$-10, -10$

11.1.1.14 Block 14

Consider the game in Table 11.6, in which player 1's strategy set is $\{A, B\}$ and player 2's strategy set is $\{C, D\}$.

1. There are multiple Nash equilibria in this game.
2. The Nash equilibrium (equilibria) is (are) Pareto-efficient.
3. The strategy profile $\{A, C\}$ Pareto-dominates the strategy profile $\{B, C\}$.
4. Not all of the best responses are unique.

11.1.2 Solutions

11.1.2.1 Sample Solutions for Block 1

See Chapter 11.5, where an almost identical game is discussed. To determine the Nash-equilibria in sequential-move games, it can be helpful to represent the game in matrix form. The payoff of a player associated with her best response has been underlined in the following matrix (Table 11.7).

1. **True**.
2. **False**.
3. **False**.
4. **True**.

11.1.2.2 Sample Solutions for Block 2

See Table 11.8. It is evident that none of the players have a dominant strategy (see Definition 11.2 in Chapter 11.3). (U, R) and (D, L) are Nash equilibria in the game.

1. **False**.
2. **False**.
3. **False**.
4. **False**.

Table 11.9 Block 3, Statement 4. A normal-form game

		Player 2			
		OO	OU	UO	UU
Player 1	Y	$\underline{1}, 0$	$\underline{1}, 0$	$\underline{1}, \underline{1}$	$\underline{1}, \underline{1}$
	X	$\underline{1}, 1$	$\underline{1}, \underline{2}$	$\underline{1}, 1$	$\underline{1}, \underline{2}$

11.1.2.3 Sample Solutions for Block 3

1. **False**. The strategy sets are $S_1 = \{Y, X\}$ for player 1 and $S_2 = \{OO, OU, UO, UU\}$ for player 2.
2. **True**. (i) Player 1 plays Y: player 2 will play U, because his utility will be 1, which is higher than 0 (the utility he would receive if he played O).
   (ii) Player 1 plays X: player 2 will play U, because his utility will be 2, which is higher than 1 (the utility he would receive if he played O).
3. **False**. Player 1 moves before player 2. Hence, it is not a simultaneous-move game. See Chapter 11.4 and Chapter 11.5.
4. **True**. It is the same game, just presented in matrix form (Table 11.9).

11.1.2.4 Sample Solutions for Block 4

1. **True**. See Chapter 11.3.
2. **True**. *"Accept"* is player B's dominant strategy. If player A is offering $Z > 0$ Swiss Francs, accepting the offer always yields a larger payoff than declining (in which case the payoff is 0). If player A offers 0 Swiss Francs, player B is indifferent between accepting and declining the offer. In this case, both, *"accept"* and *"decline"* are best responses for player B. Player A anticipates this and chooses the amount that maximizes her payoff: she offers 0 Swiss Francs.
3. **False**. Not every normal-form game has a Nash equilibrium in pure strategies but every game has at least one Nash equilibrium. See Result 11.1 in Chapter 11.4.
4. **True**. See Digression 35 in Chapter 11.5.

11.1.2.5 Sample Solutions for Block 5

1. **False**. It is not an equilibrium, because both players have an incentive to deviate from the strategy they agreed on. See Chapter 11.4.2.
2. **True**. Only the Nash equilibrium in the prisoner's dilemma is not Pareto-efficient. See Chapter 11.4.2.
3. **True**. Since there are multiple equilibria, one cannot anticipate the players' behavior solely based on the idea of the Nash equilibrium. Additionally, it also lacks a Pareto-dominant equilibrium, which could possibly help to make a prediction. See Chapter 11.4.1.
4. **True**. It is always the best response for each player to play the same strategy as the other player(s). See Chapter 11.4.1 and Definition 11.4 in Chapter 11.4.

Table 11.10 Block 6, Statements 3 and 4. A normal-form game

		Driver 2	
		L	R
Driver 1	L	10, 10	−10, −10
	R	−10, −10	10, 10

Table 11.11 Block 7. A normal-form game

		Player 2	
		L	R
Player 1	U	100, 100	50, 0
	D	100, 100	0, 150

Table 11.12 Block 8, Statement 2. A normal-form game

		Player 2	
		L	R
Player 1	U	1, 1	3, 3
	D	2, 2	4, 4

11.1.2.6 Sample Solutions for Block 6

1. **False**. If a reaction function exists for all players, then, while a Nash equilibrium will always exist, it need not be in pure strategies. See Digression 31 in Chapter 11.4.
2. **False**. If every strategy set leads to the same utility level, then any strategy combination is a Nash equilibrium. However, one cannot determine the equilibrium the players will choose (coordination problem). See Chapter 11.4.1.
3. **False**. Consider Table 11.10. Neither driver 1 nor driver 2 have a dominant strategy. See Definition 11.2 in Chapter 11.3 and Definition 11.4 in Chapter 11.4.
4. **True**. Consider Table 11.10. (L, L) and (R, R) are the Nash equilibria in this game.

11.1.2.7 Sample Solutions for Block 7

Consider the normal-form representation of the game in Table 11.11.

1. **False**. Strategy U is dominant for player 1. See Definition 11.2 in Chapter 11.3.
2. **False**. L is not player 2's best response to D.
3. **True**. (U, L) is the only Nash equilibrium in pure strategies in this game. U and L are mutual best responses.
4. **False**. D is not player 1's best response to R.

11.1.2.8 Sample Solutions for Block 8

1. **False**. Not every normal-form game has a Nash equilibrium in pure strategies. However, every game with a finite number of players and a finite number of pure strategies has at least one Nash equilibrium in mixed strategies. See Digression 31 in Chapter 11.4.
2. **True**. (D, R) is an equilibrium in dominant strategies (see Table 11.12).

Table 11.13 Block 8, Statement 3. A normal-form game

		Player 2	
		L	R
Player 1	U	$\underline{4}, \underline{4}$	3, 3
	D	2, 2	$\underline{4}, \underline{4}$

Table 11.14 Block 8, Statement 4. A normal-form game

		Player 2	
		L	R
Player 1	U	1, 1	$\underline{3}, \underline{3}$
	D	$\underline{2}, \underline{2}$	1, 1

Table 11.15 Block 9, Statement 3. A normal-form game

		Player 2			
		b_1b_3	b_1b_4	b_2b_3	b_2b_4
Player 1	a_1	$\underline{4}, \underline{4}$	$\underline{4}, \underline{4}$	2, 2	2, 2
	a_2	3, 3	$\underline{4}, \underline{4}$	$\underline{3}$, 3	$\underline{4}, \underline{4}$

Table 11.16 Block 9, Statement 4. A normal-form game

		Player 2			
		b_1b_3	b_1b_4	b_2b_3	b_2b_4
Player 1	a_1	$\underline{4}, \underline{4}$	$\underline{4}, \underline{4}$	2, 2	$\underline{2}$, 2
	a_2	3, $\underline{3}$	1, 1	$\underline{3}, \underline{3}$	1, 1

3. **False**. The game has two Nash equilibria in pure strategies, (U, L) and (D, R) (see Table 11.13).
4. **True**. (U, R) and (D, L) are Nash equilibria (see Table 11.14).

11.1.2.9 Sample Solutions for Block 9

1. **False**. Player 1's strategy set is $\{a_1, a_2\}$ and player 2's is $\{b_1b_3, b_1b_4, b_2b_3, b_2b_4\}$.
2. **True**. Every extensive-form game has at least one Nash equilibrium in pure strategies. See Digression 35 in Chapter 11.5.
3. **True**. $(a_1, b_1b_3), (a_1, b_1b_4), (a_2, b_1b_4)$, and (a_2, b_2b_4) are Nash equilibria in pure strategies (see Table 11.15).
4. **False**. There are three Nash equilibria: (a_1, b_1b_3), (a_1, b_1b_4) and (a_2, b_2b_3) (see Table 11.16).

11.1.2.10 Sample Solutions for Block 10

1. **False**. Backward induction is a concept that helps to identify non-credible strategies and to eliminate equilibria that contain empty threats. See Chapter 11.5.
2. **True**. See Definition 11.3 in Chapter 11.4.
3. **True**. If a strategy profile Pareto-dominates all other strategy profiles, then unilateral deviation from this strategy profile should make a player worse off. Hence, it has to be a Nash equilibrium. See Chapter 11.4.
4. **True**. See the discussion about focal points in Chapter 11.4.1.

11.1.2.11 Sample Solutions for Block 11

1. **False**. A strategy profile is called an equilibrium in dominant strategies if all players' strategies are dominant strategies. Equilibria in dominant strategies are thus a proper subset of all Nash equilibria, which implies that a Nash equilibrium is not necessarily also an equilibrium in dominant strategies. See Definition 11.4 in Chapter 11.4.
2. **True**. See sample solution to Block 11, Statement 1.
3. **True**. This is true by definition. See Definition 11.2 in Chapter 11.3.
4. **True**. See the discussion regarding the market-entry game in Chapter 11.5.

11.1.2.12 Sample Solutions for Block 12

1. **True**. This is true by definition. See Chapter 11.3.
2. **True**. This is true by definition. See Chapter 11.3.
3. **False**. It maximizes a player's utility for all possible strategy profiles of the other players, not just for one of them. See Definition 11.1 in Chapter 11.3.
4. **False**. Dominant-strategy equilibria exist only for a limited class of games. See Definition 11.4 in Chapter 11.4.

11.1.2.13 Sample Solutions for Block 13

1. **False**. The "dilemma" in the prisoner's dilemma is based on the fact that the strategy profile of the Nash equilibrium is the only non-Pareto-efficient strategy profile, despite the fact that both players are choosing their (dominant) best responses. See Chapter 11.4.2.
2. **True**. This is true by definition. See Chapter 11.4.3.
3. **False**. A counterexample is the prisoner's dilemma. See Chapter 11.4.2.
4. **True**. Every game in extensive form can also be portrayed in normal form. See Chapter 11.5.

11.1.2.14 Sample Solutions for Block 14

1. **True**. Both, (A, C) and (B, D), are Nash equilibria. See Chapter 11.4.1.
2. **True**. Based on both Nash equilibria, it is not possible to make at least one player better off without making another player worse off.
3. **True**. With (B, C), player 1 is better off and player 2 is not worse off than with (A, C). See Definition 11.5 in Chapter 11.4.
4. **False**. Player 1's best response to player 2 choosing C is choosing A.
   Player 1's best response to player 2 choosing D is choosing B.
   Player 2's best response to player 1 choosing A is choosing C.
   Player 2's best response to player 1 choosing B is choosing D.
   $BR_1(C) = A$, $BR_1(D) = B$, $BR_2(A) = C$, $BR_2(B) = D$.

11.2 Open Questions

11.2.1 Problems

11.2.1.1 Exercise 1

Given the following games in strategic (= normal) form, determine

- the best responses of each player,
- the Nash equilibria in pure strategies,
- the equilibria in dominant strategies,
- the Pareto-efficient Nash equilibria,
- the Pareto-dominant Nash equilibria.

For each game, draw a diagram showing the payoffs of all (pure) strategy profiles.

1. *Matching pennies* (see Table 11.17)
   The two players 1 and 2 have a penny each, which they secretly turn to heads or tails. Subsequently, they simultaneously reveal their choices. If the sides match (either both heads or both tails), player 1 wins and gets player 2's penny. If not (either heads and tails or tails and heads), player 2 wins and gets player 1's penny.
2. *The chicken game* (see Table 11.18)
   In this game, two bullies (players 1 and 2) drive their cars towards each other. Each player has to choose whether to stay on a collision course (strategy S) or to "chicken out" by turning left (strategy C).
3. *The comparative-advantage or invisible-hand game* (see Table 11.19)
   The strategies for player 1 and 2 are to either grow tomatoes (T) or pears (P).
4. *The assurance game* (see Table 11.20)
   "The process of economic development has for the most part bypassed the two hundred or so families that make up the village of Palanpur. They have remained poor, even by Indian standards. *[...]* Palanpur farmers sow their winter crops several weeks after the date at which yields would be maximized. The farmers do not doubt that earlier planting would give them larger harvest, but no one *[...]* is willing to be the first to plant, as the seeds on any alone plot would be

Table 11.17 Exercise 1.1. Matching Pennies

		Player 2	
		Heads	Tails
Player 1	Heads	$1, -1$	$-1, 1$
	Tails	$-1, 1$	$1, -1$

Table 11.18 Exercise 1.2. The chicken game

		Player 2	
		S	C
Player 1	S	$1, 1$	$8, 2$
	C	$2, 8$	$5, 5$

Table 11.19 Exercise 1.3. The invisible-hand game

		Player 2	
		P	T
Player 1	P	$2, 4$	$1, 1$
	T	$5, 5$	$4, 2$

Table 11.20 Exercise 1.4. The assurance game

		Player 2	
		E	L
Player 1	E	$4, 4$	$1, 2$
	L	$2, 1$	$3, 3$

Table 11.21 Exercise 1.5. The public-good game

		Player 2	
		N	P
Player 1	N	$2, 2$	$5, 1$
	P	$1, 5$	$3, 3$

quickly eaten by birds."[1] Assume that there are only two farmers. Their strategy sets consist of the two strategies "planting early" (E) and "planting late" (L).

5. *Public-good game or the prisoner's dilemma* (see Table 11.21)
   Two individuals, 1 and 2, living on the same street, plan to buy a street light, i.e. a local public good. The good will be provided if at least one of the two neighbors agrees to pay for the street light. If both of them agree to pay, the price of the street light will be shared equally. Both individuals decide whether to pay for the street light (P) or not (N).

11.2.1.2 Exercise 2

It is early in the morning and Mona and her flatmate Felix head back to their apartment after a night out. At home, a single bottle of beer awaits them and they do not want to store it. Both of them still being thirsty, the bottle has a value of $v = 10$ to each of them. Since they cannot agree on how to share the bottle, Mona makes the following proposition: Both have to wait until one of them gives up waiting before drinking the bottle. One minute of waiting has a value of 1, so that, for example, the net value of the bottle for Mona is 6 if Felix gives up after 4 minutes of waiting. This game can be modeled as a simultaneous move game, with the players choosing their concession time t_M and t_F at the beginning. If $t_M = t_F$, we assume that the two of them share the beer equally.

1. What is Mona's payoff function $u_M(t_M, t_F)$?
2. Find Mona's best response (BR_M) to
   a) $t_F = 5$,
   b) $t_F = 15$,
   c) $t_F = 10$.

[1] Samuel BOWLES (2003). Microeconomics: Behavior, Institutions andEvolution, Princeton: Princeton University Press, p. 23.

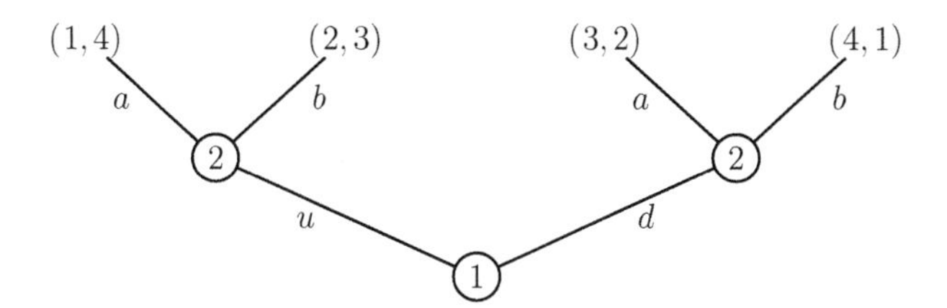

Figure 11.5 Exercise 3. A sequential game

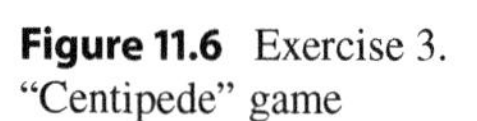
Figure 11.6 Exercise 3. "Centipede" game

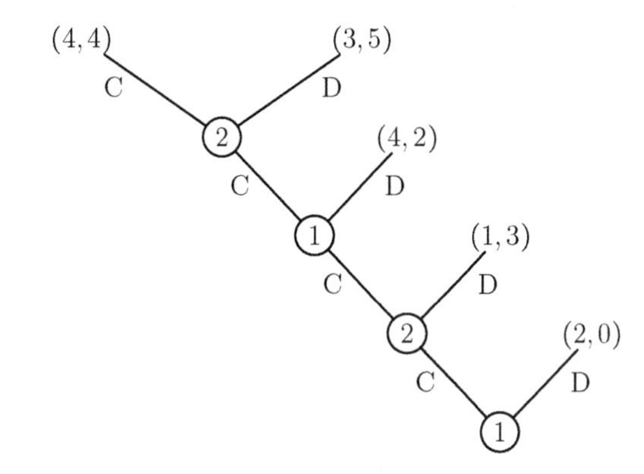

3. Do the following strategy profiles constitute a Nash equilibrium?
   a) $(t_F, t_M) = (5, 6)$,
   b) $(t_F, t_M) = (15, 0)$,
   c) $(t_F, t_M) = (10, 100)$.

11.2.1.3 Exercise 3

Given the following games in extensive form (see Fig. 11.5 and Fig. 11.6),

1. determine the players' strategy sets,
2. determine the normal-form representation of the games,
3. determine all Nash equilibria,
4. determine the number of all subgames in each game,
5. find all Nash equilibria that do not remain after backward induction, i.e. are eliminated.

11.2.2 Solutions

11.2.2.1 Solutions to Exercise 1

Some general remarks regarding Pareto efficiency and Pareto dominance:

A Nash equilibrium is

- Pareto-efficient if, given the strategy profiles, no Pareto improvement from the Nash equilibrium is possible (see Definition 5.3 in Chapter 5.1). This must be true for all possible strategy profiles of a game, not only compared to other Nash equilibria.
- Pareto-dominant if each player's utility is strictly larger in it than in all other Nash equilibria (see Definition 11.5 in Chapter 11.4.1). It follows that a Nash equilibrium can only be Pareto-dominant in a game with multiple Nash equilibria.

1. *Matching pennies* (see Table 11.22 and Fig. 11.7)
   This is a game of pure conflict, since the sum of payoffs is constant. The payoff associated with the best response of each player has been underlined in Table 11.22. For example, if player 1 chooses *Heads*, player 2's best response is *Tails*. His payoff in this case is $\underline{1}$. Figure 11.7 illustrates the payoffs associated with all possible strategy profiles.
   Obviously, there is no Nash equilibrium in pure strategies (and hence nor is there in dominant strategies), since one player always profits from unilateral deviation, regardless of the strategy profile. All strategy profiles are Pareto-efficient. Since there is no Nash equilibrium in this game, there cannot be any Pareto-dominant Nash equilibrium either.
2. *The chicken game* (see Table 11.23 and Fig. 11.8)
   There are two Nash equilibria in pure, non-dominant strategies, (S, C) and (C, S), which cannot be ranked with respect to their relative efficiency. Hence, neither Nash equilibrium Pareto-dominates the other. Both Nash equilibria as well as the strategy profile (C, C) are Pareto-efficient (see Fig. 11.8; Nash equilibria have been colored black).
3. *The comparative-advantage or the invisible-hand game* (see Table 11.24 and Fig. 11.9)
   There is a unique Nash equilibrium in pure, dominant strategies with player 1 always choosing to grow tomatoes and player 2 always choosing to grow pears. Moreover, this Nash equilibrium is the only Pareto-efficient strategy profile.
4. *The assurance game* (see Table 11.25 and Fig. 11.10)
   There are two Nash equilibria in pure, non-dominant strategies (E,E) and (L,L), where the Nash equilibrium (E,E) Pareto-dominates the Nash equilibrium (L,L). Moreover, (E, E) is the unique Pareto-efficient strategy profile.
5. *Public-good game or the prisoner's dilemma* (see Table 11.26 and Fig. 11.11)
   There is a unique Nash equilibrium in dominant strategies (N, N). There are three Pareto-efficient strategy profiles, but none of them is a Nash equilibrium.

Table 11.22 Exercise 1.1. Matching pennies

		Player 2	
		Heads	Tails
Player 1	Heads	$\underline{1}, -1$	$-1, \underline{1}$
	Tails	$-1, \underline{1}$	$\underline{1}, -1$

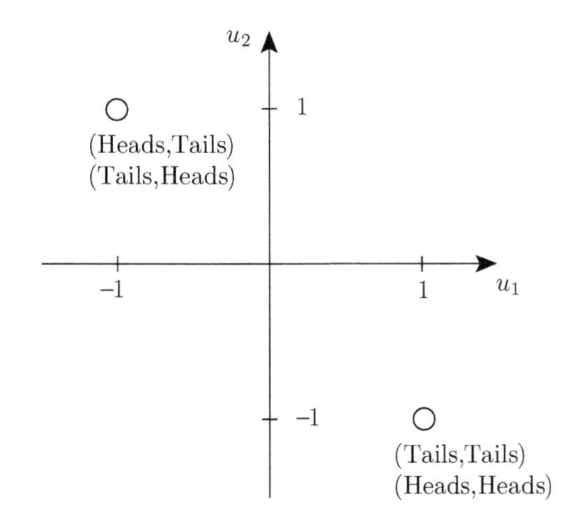

Figure 11.7 Exercise 1.1. Payoffs in the matching pennies game

Table 11.23 Exercise 1.2. The chicken game

		Player 2	
		S	C
Player 1	S	1, 1	$\underline{8}, \underline{2}$
	C	$\underline{2}, \underline{8}$	5, 5

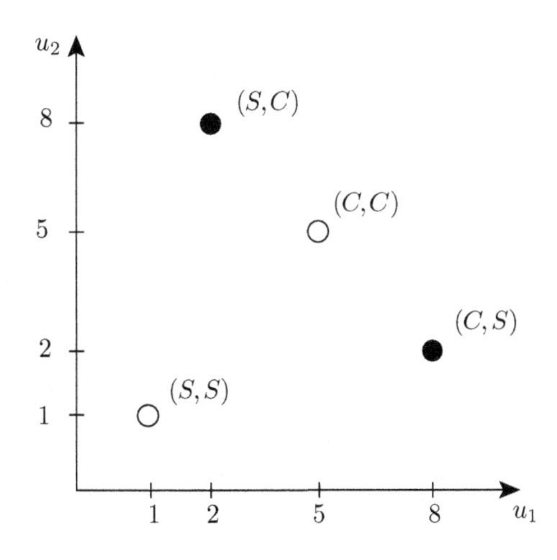

Figure 11.8 Exercise 1.2. Payoffs in the chicken game

Table 11.24 Exercise 1.3. The invisible-hand game

		Player 2	
		P	T
Player 1	P	$2, \underline{4}$	1, 1
	T	$\underline{5}, \underline{5}$	$\underline{4}, 2$

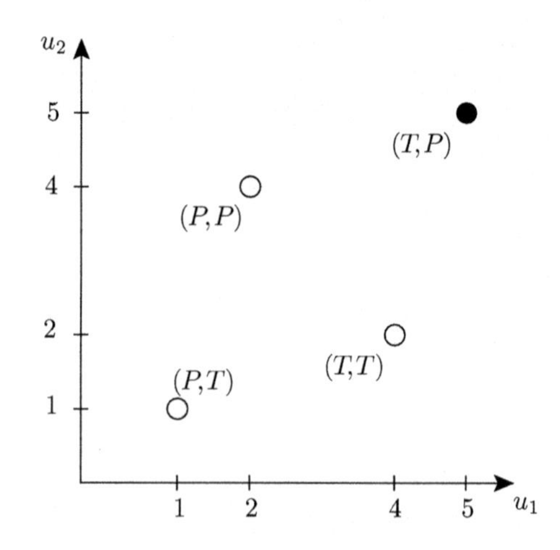

Figure 11.9 Exercise 1.3. Payoffs in the invisible-hand game

Table 11.25 Exercise 1.4. The assurance game

		Player 2	
		E	L
Player 1	E	$\underline{4}, \underline{4}$	1, 2
	L	2, 1	$\underline{3}, \underline{3}$

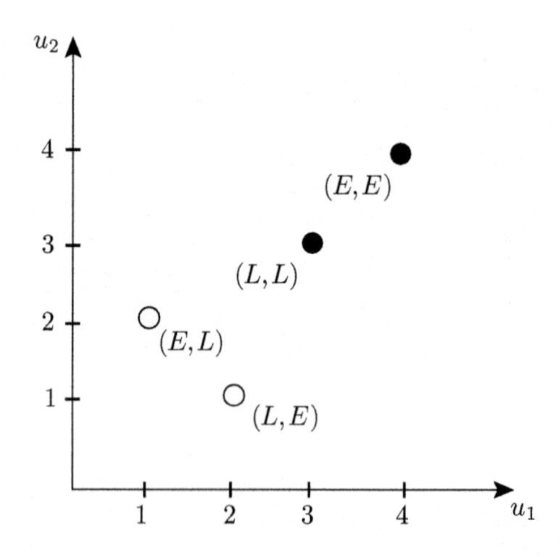

Figure 11.10 Exercise 1.4. Payoffs in the assurance game

Table 11.26 Exercise 1.5. The prisoner's dilemma

		Player 2	
		N	P
Player 1	N	$\underline{2}, \underline{2}$	$\underline{5}$, 1
	P	1, $\underline{5}$	3, 3

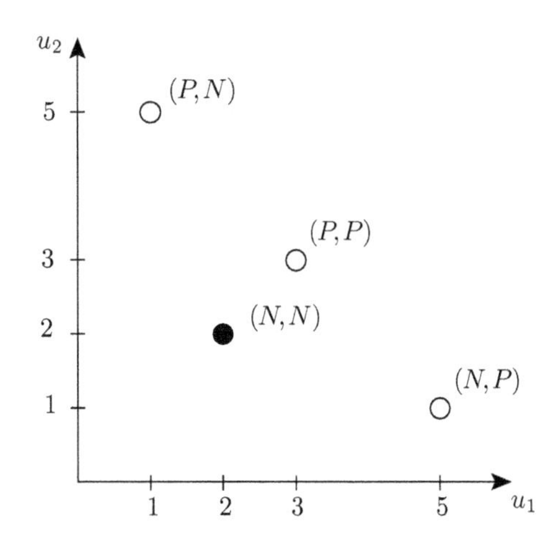

Figure 11.11 Exercise 1.5. Payoffs in the prisoner's dilemma

11.2.2.2 Solutions to Exercise 2

The situation can be modeled as a normal-form game in which each player decides, simultaneously with and independently of the opponent, the maximum time ($t_M \geq 0$, $t_F \geq 0$) that each one is willing to wait. That is, $t_F \geq 0$ is Felix's concession time, the moment he will resign if Mona does not give up first, and vice versa.

1. Mona's payoff function is thus

$$u_M(t_M, t_F) = \begin{cases} -t_M, & \text{for } t_M < t_F, \\ \frac{1}{2} \cdot 10 - t_M, & \text{for } t_M = t_F, \\ 10 - t_F, & \text{for } t_M > t_F. \end{cases}$$

2. If Mona believes that Felix' concession time is $t_F = 5$, then she is best off out-waiting Felix (since $t_F < 10$), i.e. any $t_M > 5$ is a best response for Mona (BR_M). If Mona believes that Felix's concession time is $t_F = 15$, then she is best off not waiting at all (since $t_F > 10$). Finally, for $t_F = 10$, Mona will be equally well off either conceding after $t_M = 0$ minutes, or out-waiting Felix, i.e., any $t_M > t_F$ is a best response. In general, we get:

$$BR_M(t_F) = \begin{cases} 0, & \text{for } t_F > 10; \\ > t_F, & \text{for } t_F < 10; \\ 0 \text{ and any } t_M > t_F, & \text{for } t_F = 10. \end{cases} \tag{11.1}$$

3. Requiring that each player chooses a best response to his opponent's strategy yields the following answers:

a) We already know that $t_M = 6$ is a best response for Mona to $t_F = 5$ (see Eq. 11.1). What is Felix's best response to $t_M = 6$? Since it is a symmetric game, best responses are similar to Mona's:

$$BR_F(t_M) = \begin{cases} 0, & \text{for } t_M > 10, \\ > t_M, & \text{for } t_M < 10, \\ 0 \text{ and any } t_F > t_M, & \text{if } t_M = 10. \end{cases} \tag{11.2}$$

Hence, $BR_F(t_M = 6) > 6$, and so $(t_F, t_M) = (5, 6)$ is not a Nash equilibrium.

b) We already know that $t_M = 0$ is Mona's unique best responses to $t_F = 15$. What is Felix's best response to $t_M = 0$? Looking at Eq. 11.2 we see that $BR_F(t_M = 0) > 0$. Hence, $(t_F, t_M) = (15, 0)$ is a Nash equilibrium.

c) We already know that $t_M = 100$ is a best response for Mona to $t_F = 10$. What is a best response for Felix to $t_M = 100$? Looking at Eq. 11.2 we see that $BR_F(t_M = 100) = 0$. Hence, $(t_F, t_M) = (10, 100)$ is not a Nash equilibrium.

11.2.2.3 Solutions to Exercise 3

We find the following regarding the game illustrated in Fig. 11.5.

1. Player I's strategy set is $S_{\text{I}} = \{u, d\}$, and player II's strategy set is $S_{\text{II}} = \{aa, ab, ba, bb\}$.
2. The normal-form representation is given in Table 11.27.
3. The game has two Nash equilibria in pure strategies, (d, aa) and (d, ba) (see Table 11.27).
4. The game has three subgames.
5. (d, ba) is eliminated by backward induction. (d, ba) does involve the empty promise of player II to choose b if player I chooses u.

We find the following regarding the game illustrated in Fig. 11.6.

1. Player I's strategy set is $S_{\text{I}} = \{CC, CD, DD, DC\}$, and player II's strategy set is $S_{\text{II}} = \{CC, CD, DD, DC\}$.
2. The normal-form representation is given in Table 11.28.
3. The game has four Nash equilibria in pure strategies, (DD, DD), (DD, DC), (DC, DD), and (DC, DC) (see Table 11.28).

Table 11.27 Exercise 3.2. Normal-form representation of the sequential-move game

		Player II			
		aa	ab	ba	bb
Player I	u	$1, \underline{4}$	$1, \underline{4}$	$2, 3$	$2, 3$
	d	$\underline{3}, \underline{2}$	$\underline{4}, 1$	$\underline{3}, \underline{2}$	$\underline{4}, 1$

Table 11.28 Exercise 3.2. Normal-form representation of the centipede game

		Player II			
		CC	CD	DD	DC
Player I	CC	$\underline{4}, 4$	$3, \underline{5}$	$1, 3$	$1, 3$
	CD	$\underline{4}, 2$	$\underline{4}, 2$	$1, \underline{3}$	$1, \underline{3}$
	DD	$2, \underline{0}$	$2, \underline{0}$	$\underline{2}, \underline{0}$	$\underline{2}, \underline{0}$
	DC	$2, \underline{0}$	$2, \underline{0}$	$\underline{2}, \underline{0}$	$\underline{2}, \underline{0}$

4. The game has four subgames.
5. (DD, DC), (DC, DD), and (DC, DC) are eliminated by backward induction. All of these Nash equilibria involve the non-credible promise of at least one player choosing C at least once.

11.3 Multiple Choice

11.3.1 Problems

11.3.1.1 Exercise 1

Consider the following game in extensive form (Fig. 11.12).

1. How many pure strategies do the players have?
   a) Player 1 has 2 pure strategies, player 2 has 8 pure strategies.
   b) Player 1 has 4 pure strategies, player 2 has 2 pure strategies.
   c) Player 1 has 2 pure strategies, player 2 has 4 pure strategies.
   d) Player 1 has 4 pure strategies, player 2 has 4 pure strategies.
   e) None of the above answers are correct.
2. Determine the set of Nash equilibria (in pure strategies) in this game.
   a) The set of Nash equilibria is $\{(a_1, b_1b_4)\}$.
   b) The set of Nash equilibria is $\{(a_2, b_2b_3), (a_1, b_1b_3)\}$.
   c) The set of Nash equilibria is $\{(a_1, b_1b_3), (a_2, b_2b_3), (a_1, b_2b_4)\}$.
   d) The set of Nash equilibria is $\{(a_2, b_2b_4)\}$.
   e) None of the above answers are correct.

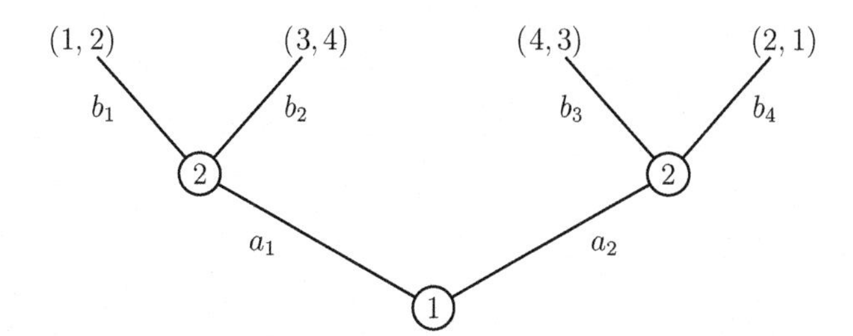

Figure 11.12 Exercise 1. A sequential game

3. Determine the dominant strategies of this game.
   a) None of the players has a dominant strategy.
   b) Player 1 has the dominant strategy a_1, player 2 has no dominant strategy.
   c) Player 1 has the dominant strategy a_1, player 2 has the dominant strategy b_2b_4.
   d) Player 1 has no dominant strategy, player 2 has the dominant strategy b_2b_3.
   e) None of the above answers are correct.
4. Determine all equilibria that do not include non-credible strategies.
   a) The set of Nash equilibria is $\{(a_1, b_1b_4), (a_1, b_2b_4)\}$.
   b) The set of Nash equilibria is $\{(a_2, b_2b_3)\}$.
   c) The set of Nash equilibria is $\{(a_1, b_2b_4)\}$.
   d) The set of Nash equilibria is $\{(a_2, b_1b_3)\}$.
   e) None of the above answers are correct.

11.3.1.2 Exercise 2

Consider the following game in normal form (see Table 11.29).

1. Determine the Nash equilibria of this game.
   a) The strategy profiles (U, L), (U, C), (M, L), and (M, C) are Nash equilibria.
   b) The strategy profiles (U, L), (U, C), (M, L), (M, C), and (D, R) are Nash equilibria.
   c) The game has no Nash equilibrium.
   d) Payoff pair $(2, 2)$ is the Nash equilibrium.
   e) None of the above answers are correct.
2. Determine the equilibria in dominant strategies of this game.
   a) The strategy profiles (U, L), (U, C), (M, L), and (M, C) are equilibria in dominant strategies.
   b) The game has no equilibria in dominant strategies.
   c) The strategy profiles (U, L), (U, C), (M, L), (M, C), and (D, R) are equilibria in dominant strategies.
   d) The equilibrium concept does not apply to this game.
   e) None of the above answers are correct.

Consider the following game in extensive form (Fig. 11.13).

3. Determine the strategies of the two players.
   a) Player 1 has strategies $\{D, U\}$. Player 2 has strategies $\{L, R\}$.
   b) Player 1 has strategies $\{D, U, L\}, \{U, D, R\}, \{D, U, R\}, \{U, D, L\}$. Player 2 has strategies $\{D, U\}$.

Table 11.29 Exercises 2.1 and 2.2. A game in normal form

		Player 2		
		L	C	R
Player 1	U	2, 2	2, 2	1, 1
	M	2, 2	2, 2	1, 1
	D	1, 1	1, 1	1, 1

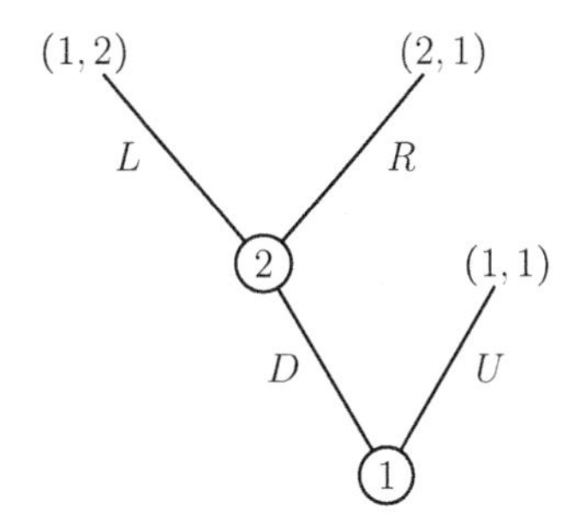

Figure 11.13 Exercises 2.3 and 2.4. A game tree

c) Player 1 has strategies $\{D, U\}$. Player 2 has strategies $\{LL, LR, RL, RR\}$.
d) Player 1 has strategies $\{D, U\}$. Player 2 has the same strategies.
e) None of the above answers are correct.

4. Determine all payoffs that result in the Nash equilibria of the game.
   a) The Nash equilibria lead to payoffs (1,1), (2,1) and (1,2).
   b) The Nash equilibria lead to payoffs (1,1).
   c) The Nash equilibria lead to payoffs (2,1) and (1,2).
   d) The Nash equilibria lead to payoffs (1,2) and (1,1).
   e) None of the above answers are correct.

11.3.1.3 Exercise 3

Consider the following game tree (Fig. 11.14):

1. What is the number of subgames of this game?
   a) It has 1 subgame.
   b) It has 2 subgames.
   c) It has 3 subgames.
   d) It has 6 subgames.
   e) None of the above answers are correct.
2. Determine player 1's and player 2's strategies.
   a) Player 1's strategy set is $\{F, S\}$ and player 2's strategy set is $\{A, D\}$.
   b) Player 1's strategy set is $\{A, D\}$ and player 2's strategy set is $\{AA, DD, AD, DA\}$.
   c) Player 1's strategy set is $\{F, S\}$ and player 2's strategy set is $\{AD, DA, DD, AA\}$.

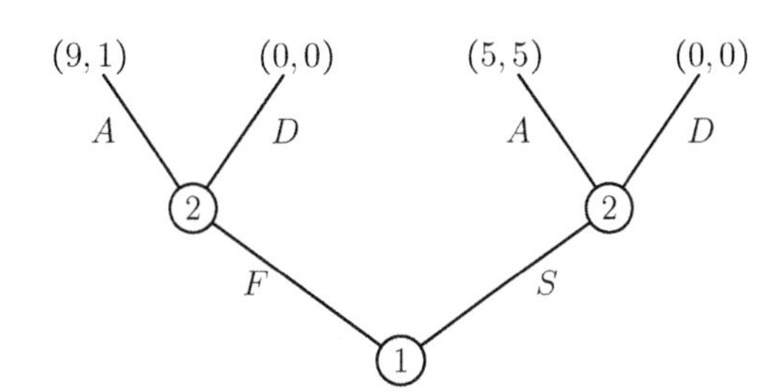

Figure 11.14 Exercises 3.1–3.4. A game tree

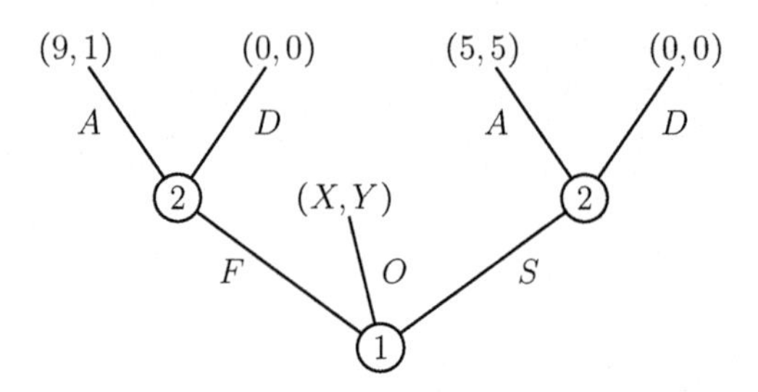

Figure 11.15 Exercises 3.5 and 3.6. A second game tree

d) Player 1's strategy set is $\{F, S\}$ and player 2's strategy set is $\{AD, AD\}$.
e) None of the above answers are correct.

3. Determine the payoffs that result for this game's Nash equilibrium (equilibria).
   a) The Nash equilibiria lead to payoffs of $(9, 1)$ and $(0, 0)$.
   b) The Nash equilibirium leads to a payoff of $(5, 5)$.
   c) The Nash equilibiria lead to payoffs of $(5, 5)$, $(5, 5)$, and $(9, 1)$.
   d) The Nash equilibiria lead to payoffs of $(9, 1)$, $(9, 1)$, and $(5, 5)$.
   e) None of the above answers are correct.
4. Determine the payoffs that result from this game when applying backward induction.
   a) The Nash equilibria lead to payoffs of $(9, 1)$ and $(0, 0)$.
   b) The Nash equilibrium leads to a payoff of $(9, 1)$.
   c) The equilibria lead to payoffs of $(5, 5)$ and $(9, 1)$.
   d) The equilibrium leads to a payoff of $(5, 5)$.
   e) None of the above answers are correct.

Consider the following extension to the game above (see Fig. 11.15).

5. Which values of $X \geq 0$ and $Y \geq 0$ would lead to four Nash equilibria with equal payoffs?
   a) $X > 9$ and $Y \geq 0$.
   b) $9 > X > 5$ and $Y > 5$.
   c) $X > 5$ and $Y \geq 0$.
   d) $X > 5$ and $9 > Y > 5$.
   e) None of the above answers are correct.
6. Which values of $X \geq 0$ and $Y \geq 0$ would lead to four Nash equilibria with two equal payoffs each?
   a) $X > 9$ and $Y > 5$.
   b) $9 > X > 5$ and $Y \geq 0$.
   c) $X > 5$ and $Y > 9$.
   d) $X > 5$ and $9 > Y > 5$.
   e) None of the above answers are correct.

11.3.2 Solutions

11.3.2.1 Solutions to Exercise 1

- Question 1, answer c) is correct.
- Question 2, answer e) is correct. The correct answer is: The Nash-equilibria are $\{(a_2, b_1b_3), (a_1, b_2b_4), (a_2, b_2b_3)\}$.
- Question 3, answer d) is correct.
- Question 4, answer b) is correct.

11.3.2.2 Solutions to Exercise 2

- Question 1, answer b) is correct.
- Question 2, answer a) is correct.
- Question 3, answer a) is correct.
- Question 4, answer d) is correct.

11.3.2.3 Solutions to Exercise 3

- Question 1, answer c) is correct.
- Question 2, answer c) is correct.
- Question 3, answer d) is correct.
- Question 4, answer b) is correct.
- Question 5, answer a) is correct.
- Question 6, answer b) is correct.

Firm Behavior in Oligopolistic Markets

12

12.1 True or False

12.1.1 Statements

12.1.1.1 Block 1

1. In a Cournot oligopoly, the firms disregard the influence of their behavior on the price.
2. In a duopoly, collusive behavior can raise overall profits.
3. In a duopoly, collusive behavior is the equilibrium strategy.
4. In a Bertrand oligopoly with symmetric firms and constant marginal costs, the equilibrium price is equal to marginal costs.

12.1.1.2 Block 2

In an oligopolistic market, all firms have identical cost functions $C(y) = c \cdot y$, with $c \geq 0$.

1. If the firms are in Bertrand price competition, there is no deadweight loss in equilibrium.
2. Collusive behavior cannot occur, because the firms have constant marginal costs.
3. In both Bertrand price competition and in Cournot quantity competition, the equilibrium market price is larger than marginal costs.
4. If the demand curve is linear and falling, the total quantity of the good supplied in Cournot competition will be lower than in Bertrand competition.

12.1.1.3 Block 3

Consider a duopoly market in which two firms produce a good with identical constant marginal costs of $MC = 0$. The demand for the total quantity y in the market is $P(y) = 300 - y$.

M. Kolmar, M. Hoffmann, *Workbook for Principles of Microeconomics*,
Springer Texts in Business and Economics, https://doi.org/10.1007/978-3-319-62662-8_12

1. In the equilibrium with Bertrand price competition, 300 units of the good are traded.
2. In the equilibrium with Cournot quantity competition, 200 units of the good are traded.
3. In equilibrium, the consumer surplus in Bertrand price competition is larger than in Cournot quantity competition.
4. The marginal costs have to be smaller than 300, in order for a positive quantity to be traded in a Cournot competition.

12.1.1.4 Block 4

1. Assume that a firm supplies a strictly positive and finite quantity. The optimality condition "marginal costs = marginal revenues" is only applicable for monopolists and not for firms in an oligopolistic market.
2. A market with Bertrand competition is served by two firms with constant and identical marginal costs. Then their revenues in the Nash equilibrium are equal to their variable costs if fixed costs are zero.
3. The inverse demand function in a duopolistic market is $P(y_1, y_2) = 90 - y_1 - y_2$. The two firms engage in Cournot competition and produce with zero marginal costs. The equilibrium price is $p^{CN} = 60$.
4. The reaction function of an oligopolist in Cournot competition informs about how much the equilibrium quantity changes if the other firms change their supply.

12.1.1.5 Block 5

1. In oligopolistic markets, few suppliers sell a homogeneous good.
2. In the Nash equilibrium of symmetric Cournot competition, the quantity supplied by a firm is smaller than the optimal supply of the same firm as a monopolist on the same market.
3. Firms in oligopolistic markets always earn zero profits in the long run.
4. The number of symmetric firms in a Cournot oligopoly increases. The market supply in the Nash equilibrium will therefore decrease.

12.1.1.6 Block 6

1. Assume a Cournot duopoly. Let firm 1's best response to some positive quantity $\hat{y}_2$ by firm 2 be zero. Therefore, $\hat{y}_2$ must then be equal to the quantity produced in optimum by a (non-price discriminating) monopolist.
2. Assume a Bertrand duopoly. A firm's best response to a price of its competitor needs not be unique.

3. Assume that the market demand function decreases linearly in the price, that two firms produce with identical constant marginal costs smaller than the consumer's maximum willingness to pay, and that fixed costs are equal to zero. In equilibrium, the consumer surplus with Bertrand competition always exceeds the consumer surplus with Cournot competition.
4. Assume a Bertrand duopoly. With identical and constant marginal costs, profits have to be equal to zero in the Nash equilibrium.

12.1.1.7 Block 7

Two firms in a Cournot oligopoly have identical and constant marginal costs, which are lower than the maximum willingness to pay in the market. The market demand function decreases linearly as the market price increases.

1. The Nash equilibrium does not have to be unique.
2. In the Nash equilibrium, both firms supply identical quantities.
3. Both firms' reaction functions do not increase as the output of the other firm increases.
4. In a Nash equilibrium, all firms have marginal revenues that are equal to marginal costs.

12.1.2 Solutions

12.1.2.1 Sample Solutions for Block 1

1. **False**. The equilibrium market price in a Cournot oligopoly is a function of the total quantity supplied. When determining the optimal strategy, a profit-maximizing Cournot oligopolist not only takes her own quantity into account, but also the quantities produced by the other firm(s). See Chapter 12.2.
2. **True**. If both firms coordinate their strategies, they can imitate the behavior of a monopolist, and split up the additional rents. See the overview table at the end of Chapter 12.
3. **False**. By deviating from collusive behavior, a firm can always increase its profits (at least in the short run). Thus, collusive behavior is not an equilibrium (see the detailed description thereof in Chapter 12.5).
4. **True**. If both firms set a price equal to marginal costs, they share the market equally, and both firms will make zero profits. Neither of the firms is better off by increasing the price (they would have no demand and their profits are still zero) nor by decreasing the price (they would serve the entire market at a price below marginal cost and thus make a loss). Consequently, setting the price equal to marginal costs is a Nash equilibrium in a Bertrand oligopoly. Moreover, it can be shown that this is the unique Nash equilibrium. See Chapter 12.4.

12.1.2.2 Sample Solutions for Block 2

1. **True**. In a Bertrand oligopoly with identical firms and constant marginal costs, the equilibrium price is equal to marginal costs. This is akin to the equilibrium under perfect competition, and there is no deadweight-loss. See Chapter 10.4 and Chapter 12.4.
2. **False**. Collusive behavior cannot occur. But the reason is not constant marginal costs. See Chapter 12.5.
3. **False**. The equilibrium price is equal to the marginal costs in Bertrand price competition.
4. **True**. To illustrate, assume a general inverse linear demand function $P(y) = a - b \cdot y$. Given identical cost functions for all n firms, the total quantity supplied in the Cournot Oligopoly is $y^{CN} = n \cdot y_i^* = \frac{n}{n+1} \cdot \frac{a-c}{b}$ (see Chapter 12.3). In a Bertrand oligopoly, "price = marginal costs" holds in equilibrium. For this cost function, this implies $a - b \cdot y = c$, or $y^B = (a - c)/b$. Therefore, it follows that:
$$y^{CN} = \underbrace{\frac{n}{n+1}}_{<1} \cdot \frac{a-c}{b} < \frac{a-c}{b} = y^B.$$

12.1.2.3 Sample Solutions for Block 3

1. **True**. Under Bertrand price competition, the equilibrium price is equivalent to the marginal costs, i.e. $p^B = 0$. Thus, $y^B = 300$. See Chapter 12.4.
2. **True**. In a duopoly with a linear supply and a linear demand function, it holds that $y^{CN} = y_1^* + y_2^* = \frac{2(a-c)}{3b}$. If $a = 300$, $b = 1$ and $c = 0$, then the result is $y^{CN} = 200$. See Chapter 12.2.
3. **True**. Given that, in equilibrium, in Bertrand price competition the price is lower and the supplied quantity is larger than in Cournot quantity competition.
4. **True**. The marginal costs of 300 correspond to the prohibitive price, i.e. $P(0) = 300$. Thus, $y^{CN} = \frac{2}{3} \frac{a-c}{b} > 0 \Leftrightarrow a > c$. See Chapter 12.3.

12.1.2.4 Sample Solutions for Block 4

1. **False**. In a Cournot oligopoly and a Bertrand oligopoly, "marginal revenues = marginal costs" is also the condition for the optimum. See Chapter 12.2.
2. **True**. In a Bertrand duopoly, "price = marginal costs" holds in equilibrium. Revenues correspond to the variable costs, given zero fixed costs and constant marginal costs. See Chapter 12.4.
3. **False**. The optimality condition for firm i is:
$$P(y_i + y_j) + \frac{\partial P(y_i + y_j)}{\partial y_i} y_i - MC(y_i) = 0,$$
$$\Leftrightarrow 90 - y_i - y_j - y_i = 0,$$

with $i \neq j$ and $i \in \{1,2\}$. Both firms' marginal costs, $MC(y_i)$, are zero. Because both firms are identical, one knows that $y_i^* = y_j^*$, in equilibrium. Thus, the conditions for the optimum are:

$$90 - y_i - y_j - y_i = 90 - 3\,y_i = 0,$$

and therefore $y_i^* = y_j^* = 30$, which leads to $p^{CN} = 30$.

4. **False**. The reaction function is a function that defines player i's best responses to the other players' possible strategy profiles. The reaction function's slope defines by how much the quantity he or she supplies changes if the quantity supplied by the other firm(s) changes. See Chapter 11.3.

12.1.2.5 Sample Solutions for Block 5

1. **True**. See Chapter 3 and Chapter 12.
2. **True**. $y_i^* = \frac{a-c}{3 \cdot b} < y^M = \frac{a-c}{2 \cdot b}$. See detailed discussion in Chapter 12.2.
3. **False**. This is not the case for the Cournot model. See Chapter 12.2.
4. **False**. In a Cournot oligopoly with several symmetrical firms, the individual supply decreases as the number of firms increases, while market supply increases. Even though all firms produce less, the effect is overcompensated as more firms produce. The market tends towards the equilibrium under perfect competition if there is a large number of firms. See Chapter 12.2.

12.1.2.6 Sample Solutions for Block 6

1. **False**. If firm 1's best response to $\hat{y}_2$ is $y_1 = 0$ firm 2 must be selling so much that the market price is equal to or below firm 1's marginal costs. In a symmetric oligopoly, this cannot be equivalent to the monopoly price. If the firms are sufficiently heterogeneous (if their marginal costs are sufficiently different), $\hat{y}_2$ can only correspond to the monopoly quantity in a rare case. See Chapter 12.2.
2. **True**. To illustrate, assume that $c_1 > 0$ are firm 1's marginal costs. Firm 1's best response to any price $p_2 < c_1$ is any price $p_1 > p_2$. See Chapter 12.4.
3. **True**. In contrast to Bertrand competition, the duopolists are able to appropriate part of the consumer surplus and make a profit with Cournot competition. See Chapter 12.4 and Chapter 12.2.
4. **True**. See Chapter 12.4.

12.1.2.7 Sample Solutions for Block 7

1. **False**. The Nash equilibrium in a symmetrical Cournot duopoly is always unique. See Chapter 12.2 and Chapter 12.3.
2. **True**. The Nash equilibrium in a symmetrical Cournot duopoly is unique and symmetrical. See Chapter 12.2 and Chapter 12.3.
3. **True**. The reaction functions are decreasing or constant in the competitor's quantity. See Chapter 12.2 and Chapter 12.3.
4. **True**. In every firm's optimum, marginal revenues equal marginal costs. See Chapter 12.

12.2 Open Questions

12.2.1 Problems

12.2.1.1 Exercise 1

Demand is given by $X(p) = 100 - p$. There are two firms in the market, each producing the good with constant marginal costs of $c = 10$ and no fixed costs. Both firms engage in Bertrand competition, i.e. the firms choose prices p_1 and p_2. If prices are equal, both producers share the demand equally.

1. Determine the firm-specific demand of firm 1 contingent on p_1 and p_2, i.e. $x_1(p_1, p_2)$.
2. Draw firm 1's profit function contingent on $p_1 \geq 0$ under the following conditions:
   a) $0 < p_2 < 10$,
   b) $p^M < p_2$,
   c) $p_2 = 10$,
   d) $10 < p_2 \leq p^M$,
   where p^M represents the monopoly price.
3. Determine the reaction function $P_1(p_2)$ of firm 1.
4. Determine the price (p^B) and quantity (x^B) at the Nash equilibrium.
5. Determine the producer surplus $PS(x^B)$ and the consumer surplus $CS(x^B)$ at the Nash equilibrium.
6. Is the market equilibrium Pareto-efficient?

12.2.1.2 Exercise 2

The inverse demand function in a Cournot duopoly is given by $P(y) = 100 - y$. The cost function of the firms in the market is given by $C_1(y_1) = c\, y_1$ and $C_2(y_2) = c\, y_2$, with $c = 10$. Supply of firm i is denoted by y_i, with $i \in \{1, 2\}$. Total supply is equal to $y_1 + y_2 = y$.

1. Draw the profit function of firm 1 contingent on y_1, given the following values of y_2:
   a) $y_2 = 20$,
   b) $y_2 = 30$.
   Comment on your findings.
2. Determine the reaction function $Y_1(y_2)$ of firm 1. What is the reaction $Y_1(y_2)$ of firm 1 to
   a) $y_2 = 0$,
   b) $y_2 = 90$?
   Comment on your findings.
3. Draw both firm's reaction functions in the strategy space (y_2, y_1) and highlight the Nash equilibrium.
4. Determine the price (p^{CN}) and market supply (y^{CN}) at the Nash equilibrium.

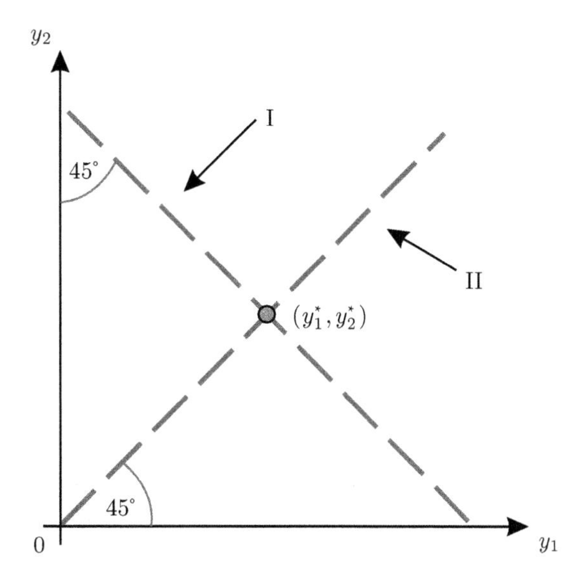

Figure 12.1 Exercise 2.7. Nash equilibrium in a Cournot duopoly

5. Determine the producer surplus $PS(y^{CN})$ and the consumer surplus $CS(y^{CN})$ at the Nash equilibrium.
6. Is the market equilibrium Pareto-efficient?
7. Consider Fig. 12.1. Suppose that, starting from the Nash equilibrium in the Cournot duopoly (y^{CN}), firms can only move along the lines I and II. In which direction should the two firms move in order to increase both their profits? What has to be taken into consideration when doing so?

Please consider the general case with n identical firms. Market demand and cost functions are the same as above.

8. Calculate the individual and market supply as well as the equilibrium price in the Nash equilibrium.
9. How does the number of firms (n) influence individual (y_i^*) and market supply (y^{CN}) as well as market price (p^{CN}) in the Nash equilibrium?

12.2.2 Solutions

12.2.2.1 Solutions to Exercise 1

1. If $p_1 < p_2$, firm 1 is the unique supplier and, thus firm-specific demand equals $X(p_1)$. For $p_1 = p_2$, both firms share the market equally, for $p_1 > p_2$, the

firm-specific demand of the first firm is zero. We thus get:

$$x_1(p_1, p_2) = \begin{cases} 0, & \text{for } p_1 > p_2, \\ \frac{1}{2}X(p_1) & \text{for } p_1 = p_2, \\ X(p_1) & \text{for } p_1 < p_2, \end{cases} = \begin{cases} 0, & \text{for } p_1 > p_2, \\ 50 - \frac{1}{2}\, p_1 & \text{for } p_1 = p_2, \\ 100 - p_1 & \text{for } p_1 < p_2. \end{cases} \quad (12.1)$$

2. The profit function of firm 1 is

$$\pi_1(p_1, p_2) = \underbrace{p_1 \cdot x_1(p_1, p_2)}_{=\text{ revenues}} - \underbrace{c \cdot x_1(p_1, p_2)}_{=\text{ costs}} = (p_1 - 10) \cdot x_1(p_1, p_2).$$

Assume that $p_1 < p_2$, then $x_1(p_1, p_2) = X(p_1) = 100 - p_1$ (see Eq. 12.1). The profit function of firm 1 is then given by

$$\pi_1(p_1, p_2) = p_1 \cdot X(p_1) - c \cdot X(p_1) = (p_1 - 10) \cdot (100 - p_1).$$

Taking the first and second derivative with respect to p_1 yields

$$\frac{\partial \pi_1(p_1, p_2)}{\partial p_1} = 110 - 2\, p_1,$$
$$\frac{\partial^2 \pi_1(p_1, p_2)}{\partial p_1^2} = -2.$$

Thus, if we assume that $p_2 > p_1$, we find that $\pi_1(p_1, p_2)$ is a strictly concave function, with a global maximum at

$$\frac{\partial \pi_1(p_1, p_2)}{\partial p_1} = 0 \quad \Leftrightarrow \quad 110 - 2p_1 = 0 \quad \Leftrightarrow \quad p_1 = 55,$$

see Fig. 12.2.

Moreover, we know that $\pi_1(0, p_2) = -c \cdot X(0) = -10 \cdot 100 = -1{,}000$ for $p_2 > 0$. This means that at $p_2 > 0$ and $p_1 = 0$, profits of firm 1 equal $-1{,}000$ Swiss Francs for the following reason: Since $p_1 = 0 < p_2$, the demand firm 1 is facing is the total demand in the market. At price $p_1 = 0$, total demand equals 100 units of the good. Firm 1 has constant marginal costs and thus also constant average variable costs $c = AVC = 10$, which leads to total profits of $-1{,}000$.

a) In this case, the profit increases in p_1 (but stays negative) until we reach $p_1 = p_2$ (see Fig. 12.3). At $p_1 = p_2$ the profit function is discontinuous because both firms share the market (see Eq. 12.1) and therefore, also share the loss. Profit is $\pi_1(p_1, p_2) = p_1 \cdot (50 - \frac{1}{2}\, p_1) - 10 \cdot (50 - \frac{1}{2}\, p_1) = (p_1 - 10) \cdot (50 - \frac{1}{2}\, p_1)$ at $p_1 = p_2$. For $p_1 > p_2$, profit is zero.

b) We already know that the profit function has a global maximum at $p_1 = 55$ for $p_1 < p_2$ (see Fig. 12.2). Thus, $p^M = 55$. We assumed that $p_2 > p^M$. In this case, $\pi_1(p_1, p_2)$ increases in p_1 until $p_1 = 55$ (see Fig. 12.4). Profit decreases until $p_1 = p_2$, where the profit function is discontinuous, because both firms share the market and therefore, also the profit. For $p_1 > p_2$, we get $\pi_1(p_1, p_2) = 0$.

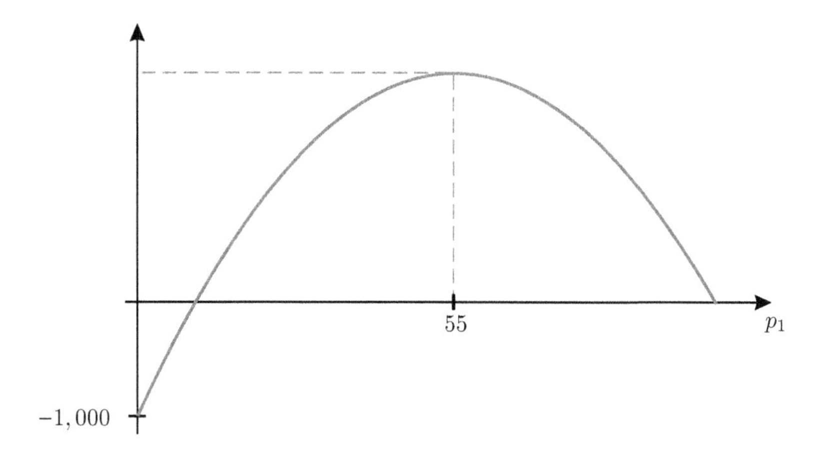

Figure 12.2 Exercise 1.2. $\pi_1(p_1, p_2)$ for $p_1 < p_2$

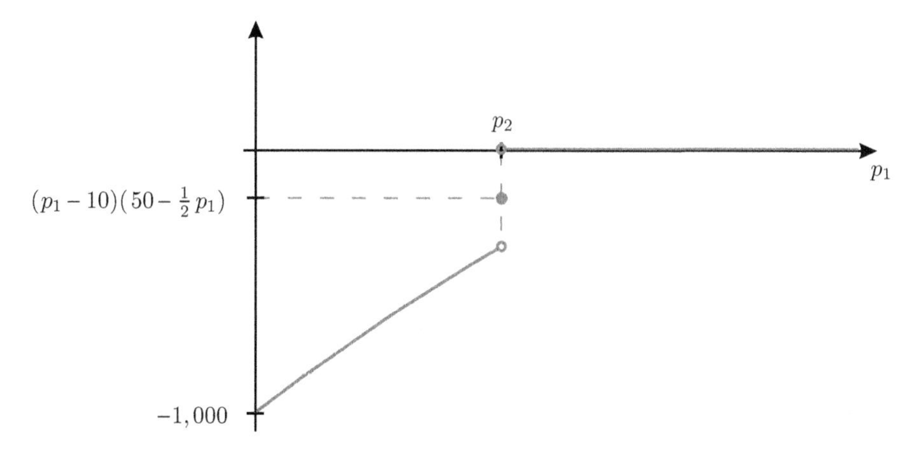

Figure 12.3 Exercise 1.2a). $\pi_1(p_1, p_2)$ for $0 < p_2 < 10$

c) This case is similar to the one discussed in Question 2a). The only difference is that at $p_1 = p_2 = 10$ the discontinuity of the profit function vanishes. The reason is that profit is zero at $p_1 = p_2$ (since the price equals the marginal costs), and this holds for all values of $p_1 \geq p_2$ (see Fig. 12.5).

d) Since $p^M \geq p_2 > c = 10$, firm 1's profit is positive for $p_2 > p_1 > c = 10$ and increases in p_1 until $p_1 = p_2$ (see Fig. 12.6). Then again, profit is zero for $p_1 > p_2$.

3. Next, we derive the best response of firm 1: For $p_2 < c$, any price $p_1 \leq p_2$ leads to a negative profit, while any price $p_1 > p_2$ leads to zero profit (see Question 2a)). Thus, $P_1(p_2)$ is any p_1 larger than p_2. For $p_2 = c$ we need a slight change in the argument: We find that firm 1's profit is negative as long as $p_1 < p_2$, whereas it becomes zero for all $p_1 \geq p_2$ (see Question 2c)). Hence, any price

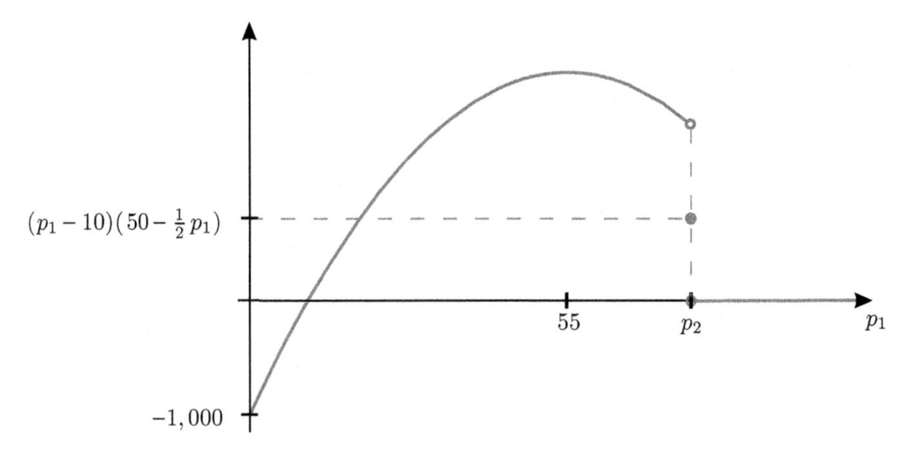

Figure 12.4 Exercise 1.2b). $\pi_1(p_1, p_2)$ for $p^M < p_2$

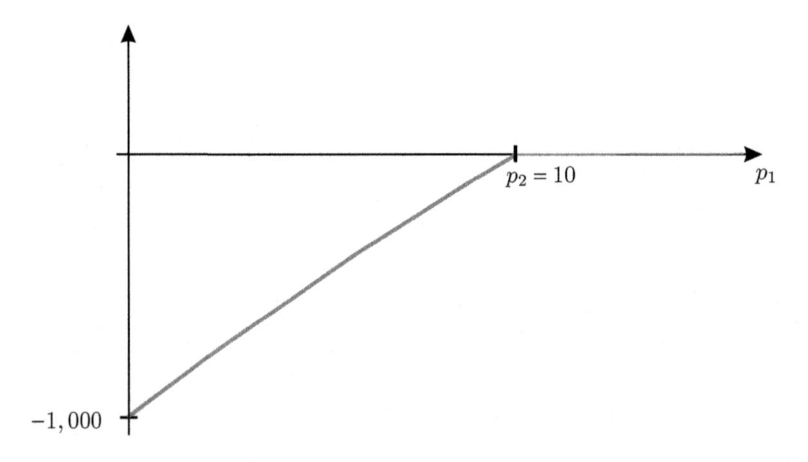

Figure 12.5 Exercise 1.2c). $\pi_1(p_1, p_2)$ for $p_2 = c$

larger or equal to p_2 is a best response. For $c < p_2 \leq p^M$, profit increases as long as $p_1 < p_2$ (see Question 2d)). Firm 1's best response is therefore to undercut firm 2's price slightly. For $p_2 > p^M$, the best response is the monopoly price ($P_1(p_2) = p^M$, see Question 2b)). The reaction function of firm 1 is thus

$$P_1(p_2) = \begin{cases} > p_2, & \text{for } p_2 < 10, \\ \geq p_2, & \text{for } p_2 = 10, \\ p_2 - \varepsilon, & \text{for } 10 < p_2 \leq 55 = p^M, \\ 55, & \text{for } p_2 > 55 = p^M, \end{cases} \tag{12.2}$$

with $\varepsilon > 0$, $\varepsilon \to 0$.

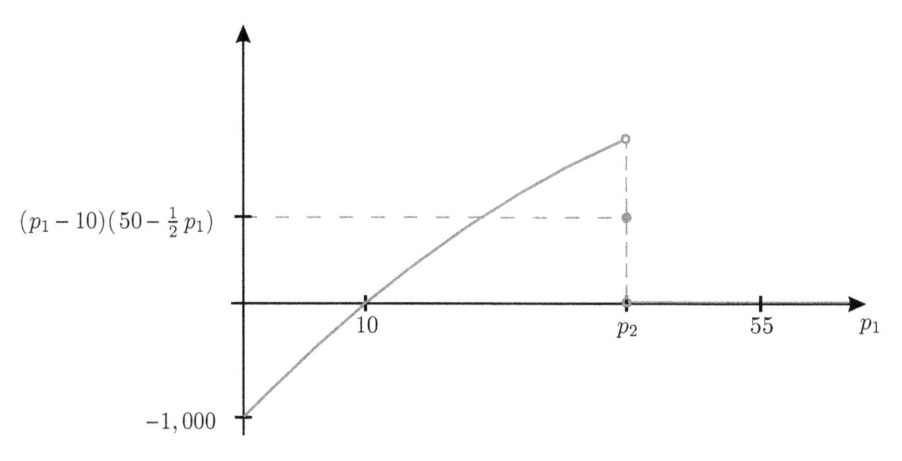

Figure 12.6 Exercise 1.2d). $\pi_1(p_1, p_2)$ for $c < p_2 < p^M$

4. In a Nash equilibrium none of the firms are able to improve upon the profit at (p_1^*, p_2^*) by unilateral deviation. Or, to put it differently, each equilibrium price is a best response to the strategy chosen by the opponent. Since we have a symmetric game, firm 2's reaction function $P_2(p_1)$ can be obtained by replacing p_2 with p_1 in Eq. 12.2. Obviously, there is only one strategy profile such that no firm has an incentive to deviate: $p_1^* = p_2^* = p^B = 10$. The corresponding quantity is then given by $x^B = \frac{1}{2}X(p_1^*) + \frac{1}{2}X(p_2^*) = 90$ (see Eq. 12.1).
5. Since the equilibrium price equals the marginal costs, we find that $PS = 0$, whereas
$$CS(x^B) = (100 - 10) \cdot X(10) \cdot \frac{1}{2} = 4{,}050.$$
6. No other strategy profile (p_1, p_2) can increase the sum of consumer and producer surplus above 4,050. Hence, the market equilibrium is Pareto-efficient.

12.2.2.2 Solutions to Exercise 2

1. The profit function of firm 1 is given by
$$\pi_1(y_1, y_2) = \underbrace{P(y_1, y_2) \cdot y_1}_{= \text{revenues}} - \underbrace{C_1(y_1)}_{= \text{costs}} = (100 - \underbrace{(y_1 + y_2)}_{= y}) \cdot y_1 - 10 \cdot y_1.$$
We find that
$$\frac{\partial \pi_1(y_1, y_2)}{\partial y_1} = \underbrace{100 - 2\,y_1 - y_2}_{= \text{marginal revenues}} - \underbrace{10}_{= \text{marginal costs}},$$
$$\frac{\partial^2 \pi_1(y_1, y_2)}{\partial y_1^2} = -2.$$

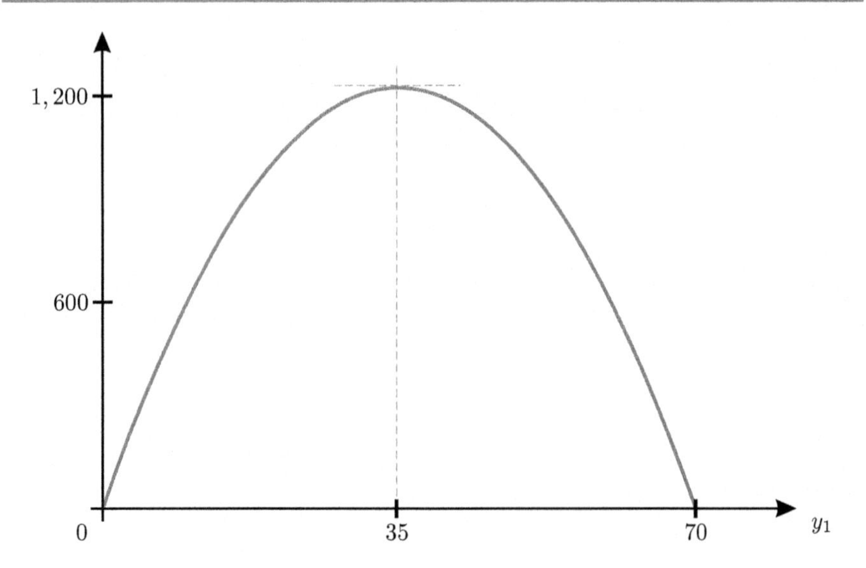

Figure 12.7 Exercise 2.1a). $\pi_1(y_1, 20)$

Thus, $\pi_1(y_1, y_2)$ is a strictly concave function with a maximum at

$$\frac{\partial \pi_1(y_1, y_2)}{\partial y_1} = 0 \quad \Leftrightarrow \quad 90 - 2\,y_1 - y_2 = 0 \quad \Leftrightarrow \quad \tilde{y}_1 = 45 - \frac{1}{2} y_2. \quad (12.3)$$

Moreover, we know that

$$\begin{aligned} \pi_1(y_1, y_2) = 0 \quad &\Leftrightarrow \quad P(y_1, y_2)\, y_1 = c\, y_1 \\ \Rightarrow \quad y_1 = 0 \quad &\vee \quad P(y_1, y_2) = c. \end{aligned}$$

In words, the profit of firm 1 is zero if either $y_1 = 0$, or if the total supply is large enough so that the market price equals marginal costs.

a) Given $y_2 = 20$, the profit function is displayed in Fig. 12.7. The function is strictly concave with a maximum at $\tilde{y}_1 = 45 - \frac{1}{2} \cdot 20 = 35$. How does it alter the shape of the profit function if we change the supply of the other firm (y_2)?
b) First, we need to analyze how $\pi_1(y_1, y_2)$ reacts to an increase in y_2:

$$\frac{\partial \pi_1(y_1, y_2)}{\partial y_2} = -y_1.$$

Thus, firm 1's profit decreases in y_2 if $y_1 > 0$. This is in accordance with economic intuition: The larger the opponent's supply, the smaller, *ceteris paribus*, the market price, given the inverse demand function $P(y_1, y_2)$. Consequently, profit also decreases. In terms of Fig. 12.7, we thus know

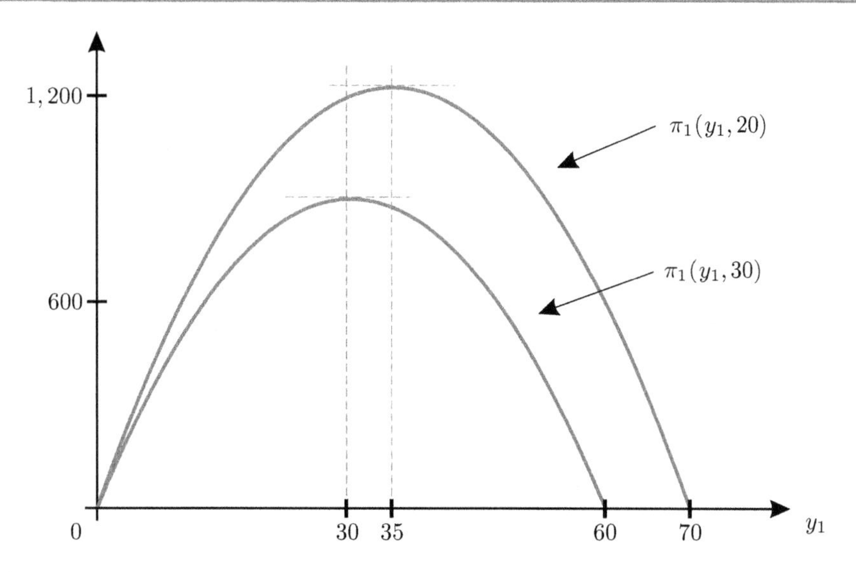

Figure 12.8 Exercise 2.1b). $\pi_1(y_1, y_2)$ for $y_2 = 20$ and $y_2 = 30$

that the course of $\pi_1(y_1, 30)$ lies below $\pi_1(y_1, 20)$. However, we still do not know how the profit-maximizing level of y_1 (i.e., $\tilde{y}_1$) changes due to the increase in y_2.
Utilizing Eq. 12.3, we find that $\tilde{y}_1 = 45 - \frac{1}{2}y_2 = 45 - \frac{1}{2} \cdot 30 = 30$ (see Fig. 12.8). The larger the opponent's supply, the lower the profit-maximizing level of a firm's own supply.

2. The reaction function of firm 1 gives us the profit-maximizing level of y_1 for all possible values of $y_2 \geq 0$. We have already determined that value ($\tilde{y}_1$, see Eq. 12.3). However, this value can become negative. We thus have to distinguish between the following cases:

$$Y_1(y_2) = \begin{cases} \tilde{y}_1, & \text{for } y_2 \leq 90, \\ 0, & \text{else,} \end{cases} = \begin{cases} 45 - \frac{1}{2}y_2, & \text{for } y_2 \leq 90, \\ 0, & \text{else.} \end{cases} \tag{12.4}$$

a) At $y_2 = 0$ we find that $Y_1(0) = 45$. This is equal to the supply that a profit-maximizing monopolist would offer, since at $y_2 = 0$, firm 1 acts as a monopolist.
b) At $y_2 = 90$ we find that $Y_1(90) = 0$. In this case, the opponent's supply is so large that $P(y_1, 90) = 10 - y_1$. Hence, at $y_1 = 0$ the market price equals the marginal costs of firm 1, $P(0, 90) = 10$. For $y_1 > 0$ it would even fall below that value. Consequently, that firm's best response is $y_1 = 0$. Therefore, the total supply equals the supply of firm 2 and $y_1 + y_2 = 90$. Since both firms have identical and constant marginal costs, supply is equal to the amount supplied in a market with perfect competition.

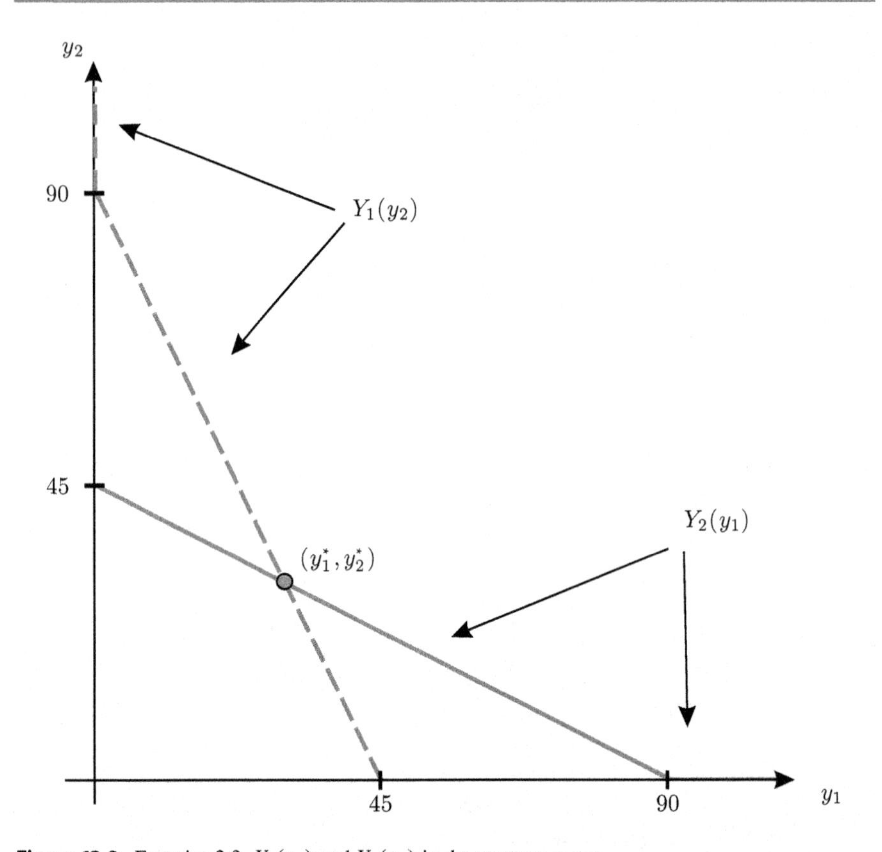

Figure 12.9 Exercise 2.3. $Y_1(y_2)$ and $Y_2(y_1)$ in the strategy space

3. The shape of both reaction functions is illustrated in Fig. 12.9. In the Nash equilibrium (y_1^*, y_2^*), both strategies are mutually best responses, i.e. $y_1^* = Y_1(y_2^*)$ and $y_2^* = Y_2(y_1^*)$. This holds at the (unique) intersection of both reaction functions.
4. We know that for any strategy profile on the reaction function of player 1 we get:

$$y_1 = Y_1(y_2).$$

At the Nash equilibrium, we additionally have that y_2 is a best response to y_1. Consequently, we get

$$y_1^* = Y_1(\underbrace{Y_2(y_1^*)}_{=y_2^*}),$$

$$\Leftrightarrow \quad y_1^* = 45 - \frac{1}{2}\underbrace{\left(45 - \frac{1}{2}y_1^*\right)}_{=y_2^*}$$

$$\Leftrightarrow \quad y_1^* = 22.5 + \frac{1}{4} y_1^*$$

$$\Leftrightarrow \quad y_1^* = 30.$$

Hence, $y_1^* = 30$, and (since we have a symmetric oligopoly) $y_2^* = 30$. Total supply equals $y^{CN} = y_1^* + y_2^* = 60$, and the market price equals $p^{CN} = P(y_1^*, y_2^*) = 100 - 30 - 30 = 40$.

5. Producer surplus equals the sum of both firm's profits at (y_1^*, y_2^*). We start by determining the equilibrium profit of one firm:

$$\pi_1^* = \pi_1(y_1^*, y_2^*) = y_1^* \cdot p^{CN} - c \cdot y_1^* = (p^{CN} - c) \cdot y_1^* = (40 - 10) \cdot 30 = 900.$$

Hence,

$$PS(y^{CN}) = \pi_1^* + \pi_2^* = 1{,}800.$$

The consumer surplus at $p^{CN} = 40$ is

$$CS(y^{CN}) = (100 - 40) \cdot 60 \cdot \frac{1}{2} = 1{,}800.$$

6. No. Since $p^{CN} > c$, total supply can be increased until $p = c$.
7. a) If both firms move along line I, then
   (i) total supply in the market remains constant, i.e. on line I it is true that $y_1 + y_2 = y^{CN}$,
   (ii) the quantity supplied by one firm increases, while the quantity supplied by the other firm decreases by the exact same amount.

   Therefore, moving along line I has no impact on the price as $p^{CN} = P(y^{CN})$.

   However, profits are redistributed from one firm to the other, while the sum of profits remains constant.

   b) If both firms move along line II in a northeastern direction, then
   (i) total supply in the market increases while
   (ii) the profit shares of both firms do not change.

   Since the quantity supplied in the Nash equilibrium of the Cournot duopoly ($y^{CN} = 60$) is already larger than the quantity supplied in monopoly ($y^M = 45$, see Question 2a)), a further increase in the supplied quantity will reduce total profits.

   c) Moving along line II in a southwestern direction does not change the profit shares of both firms, but reduces the total quantity supplied in the market. Thus, this has to increase the profit of both firms. It has to be taken into consideration that the movement along line II does not impact profits monotonically. To illustrate this, note that the profit of both firms in point $y_1 + y_2 = 0$ is smaller than the profit in the Cournot Nash equilibrium.

8. Similar to Question 1, the profit function of firm i is defined as:

$$\pi_i(y_i, Y_{-i}) = \underbrace{P(y_i, Y_{-i}) \cdot y_i}_{= \text{revenues}} - \underbrace{C(y_i)}_{= \text{costs}} = (100 - y_i - Y_{-i}) \cdot y_i - 10\, y_i,$$

where $Y_{-i} = y_1, \ldots, y_{i-1}, y_{i+1}, \ldots, y_n$. It follows that

$$\frac{\partial \pi_i(y_i, Y_{-i})}{\partial y_i} = \underbrace{100 - 2\, y_i - Y_{-i}}_{= \text{marginal revenues}} - \underbrace{10}_{= \text{marginal costs}},$$

$$\frac{\partial^2 \pi_i(y_i, Y_{-i})}{\partial y_i^2} = -2.$$

The profit function $\pi_i(y_i, Y_{-i})$ is a strictly concave function, and its maximum is determined by

$$\frac{\partial \pi_i(y_i, Y_{-i})}{\partial y_i} = 0 \quad \Leftrightarrow \quad 90 - 2\, y_i - Y_{-i} = 0. \tag{12.5}$$

Remember that all firms are identical. Thus, all n firms will supply the same quantity y^* at the Nash equilibrium, i.e. $y_i = y^*$ and $Y_{-i} = (n-1) \cdot y^*$. It follows from Eq. 12.5 that

$$\begin{aligned} \left.\frac{\partial \pi_i(y_i, Y_{-i})}{\partial y_i}\right|_{y_i = y^*, Y_{-i} = (n-1)\cdot y^*} &= 0 \\ \Leftrightarrow 90 - 2y^* - (n-1) \cdot y^* &= 0 \\ \Leftrightarrow 90 &= (n+1)\, y^* \\ \Leftrightarrow y^* &= \frac{90}{n+1}. \end{aligned} \tag{12.6}$$

This yields the following market supply in equilibrium:

$$y^{CN} = n \cdot y^* = 90 \cdot \frac{n}{n+1}. \tag{12.7}$$

Utilizing the inverse demand function yields:

$$p^{CN} = 100 - y^{CN} = 100 - 90 \cdot \frac{n}{n+1}. \tag{12.8}$$

9. a) The individual supply of a firm i decreases in the number n of firms in the market. It follows from Eq. 12.6:

$$\frac{d\, y^*}{d\, n} = -\frac{90}{(n+1)^2} < 0.$$

b) The market supply y^{CN} increases in the number n of firms in the market. It follows from Eq. 12.7:

$$\frac{d\,y^{CN}}{d\,n} = \frac{90}{(n+1)^2} > 0.$$

c) Additionally, consider the relationship between market price p^{CN} and the number of firms n. The market price decreases in the number of firms. It follows from Eq. 12.8:

$$\frac{d\,p^{CN}}{d\,n} = -\frac{90}{(n+1)^2} < 0.$$

Given identical firms, we can conclude that the larger the number of firms, the more competitive and efficient the market.

12.3 Multiple Choice

12.3.1 Problems

12.3.1.1 Exercise 1

Two identical firms have the following production function: $Y(l) = 5\,l$. The wage is $w = 10$, and there are no fixed costs. The inverse market demand for the good is $P(y) = 95 - y$.

1. Determine each firm's quantity in the Nash equilibrium under Bertrand competition.
   a) Indeterminate.
   b) $y_i^* = \frac{93}{2}$.
   c) $y_i^* = \frac{237}{5}$.
   d) $y_i^* = \frac{45}{2}$.
   e) None of the above answers are correct.
2. Determine the market price in the Nash equilibrium under Bertrand competition.
   a) Indeterminate.
   b) $p^B = 2$.
   c) $p^B = \frac{1}{5}$.
   d) $p^B = 50$.
   e) None of the above answers are correct.
3. Determine each firm's quantity in the Nash equilibrium under Cournot competition.
   a) Indeterminate.
   b) $y_i^* = 31$.
   c) $y_i^* = \frac{93}{2}$.
   d) $y_i^* = 15$.
   e) None of the above answers are correct.

4. Determine the market price in the Nash equilibrium under Cournot competition.
   a) Indeterminate.
   b) $p^{CN} = 64$.
   c) $p^{CN} = 65$.
   d) $p^{CN} = 33$.
   e) None of the above answers are correct.

12.3.1.2 Exercise 2

In a goods market, there is the following inverse market demand function $P(y) = \frac{1}{y} - 1$, where $y = y_1 + y_2$. Two duopolists $i = 1, 2$ can supply this market without marginal or fixed costs.

1. Calculate the quantity and the price in a Nash equilibrium under Cournot competition.
   a) $p^{CN} = 0$ and $y_i^* = 0.25$ for each firm.
   b) $p^{CN} = 1$ and $y_i^* = 0.25$ for each firm.
   c) $p^{CN} = 3$ and $y_i^* = 0.125$ for each firm.
   d) The equilibrium price and the equilibrium quantity are indeterminate.
   e) None of the above answers are correct.
2. Calculate the quantity and the price in the Nash equilibrium in this market under Bertrand competition.
   a) $p^B = 0$ and $y_i^* = 0.5$ for each firm.
   b) $p^B = 1$ and $y_i^* = 0.25$ for each firm.
   c) $p^B \to \infty$ and $y_i^* \to \infty$ for each firm, where the arrows indicate the convergence of price and quantity towards the annotated amount.
   d) The equilibrium price and the equilibrium quantity are indeterminate.
   e) None of the above answers are correct.

12.3.1.3 Exercise 3

Assume two firms, 1 and 2, produce a homogeneous good with identical constant marginal costs of 25 and without fixed costs. The inverse demand is given by $P(y) = 145 - y$.

1. How much do the firms supply and what is the price in the Nash equilibrium under Bertrand competition?
   a) The price is $p^B = 50$, the quantity is $y^B = 95$.
   b) The price is $p^B = 25$, the quantity is $y^B = 120$.
   c) The price is $p^B = 51$, the quantity is $y^B = 94$.
   d) The price is $p^B = 26$, the quantity is $y^B = 119$.
   e) None of the above answers are correct.
2. What is the deadweight loss (DWL) of the Nash equilibrium under Bertrand competition compared to the Pareto optimum?
   a) The deadweight loss is 240.
   b) The deadweight loss is 340.
   c) The deadweight loss is 360.

d) The deadweight loss is 0.
e) None of the above answers are correct.

3. How much do the firms supply and what is the price in the Nash equilibrium under Cournot competition?
   a) The price is $p^{CN} = 25$, the quantity is $y^{CN} = 120$.
   b) The price is $p^{CN} = 55$, the quantity is $y^{CN} = 90$.
   c) The price is $p^{CN} = 65$, the quantity is $y^{CN} = 80$.
   d) The price is $p^{CN} = 45$, the quantity is $y^{CN} = 100$.
   e) None of the above answers are correct.
4. What is the deadweight loss (DWL) in the Nash equilibrium under Cournot competition compared to the Pareto optimum?
   a) The deadweight loss is 2,300.
   b) The deadweight loss is 800.
   c) The deadweight loss is 1,600.
   d) The deadweight loss is 825.
   e) None of the above answers are correct.
5. The firms are in a Cournot-quantity competition. A regulation agency regulates the firms in order to increase welfare. Due to the agency's intervention, the firms' competitive behavior changes and Bertrand-price competition ensues. The bureau causes direct social costs of C. How large can C be at maximum, so that the existence of the agency increases welfare, when taking these costs into consideration?
   a) C cannot be larger than the sum of consumer and producer surplus in the Nash equilibrium under Bertrand competition.
   b) C cannot be larger than the deadweight loss in the Nash equilibrium under Cournot competition.
   c) C cannot be larger than the sum of consumer and producer surplus in the Nash equilibrium under Cournot competition.
   d) The amount of C is irrelevant to the total welfare in oligopolistic markets.
   e) None of the above answers are correct.

12.3.1.4 Exercise 4

There are two firms, $i = 1, 2$, in a market. Their cost functions are $C_i(y_i) = 9\, y_i$. The inverse demand function is $P(y) = 201 - y$, with $y = y_1 + y_2$.

1. Determine the quantity and the price in the Nash equilibrium under Cournot competition.
   a) The quantity is $y_1^* = y_2^* = 64$, the price is $p^{CN} = 73$.
   b) The quantity is $y_1^* = y_2^* = 50$, the price is $p^{CN} = 101$.
   c) The quantity is $y_1^* = y_2^* = 75$, the price is $p^{CN} = 51$.
   d) The quantity is $y_1^* = y_2^* = 79$, the price is $p^{CN} = 43$.
   e) None of the above answers are correct.

2. Determine the quantity and the price in the Nash equilibrium under Bertrand competition.
   a) The total quantity is $y^B = 190$, the price is $p^B = 11$.
   b) The total quantity is $y^B = 150$, the price is $p^B = 52$.
   c) The total quantity is $y^B = 192$, the price is $p^B = 9$.
   d) The total quantity is $y^B = 170$, the price is $p^B = 31$.
   e) None of the above answers are correct.

Assume that a third firm enters the market with the same cost function.

3. How do the total supply in the market, the market price, and welfare change in the Nash equilibrium under Cournot competition?
   a) Total supply increases, the price decreases, welfare decreases.
   b) Total supply decreases, the price decreases, welfare increases.
   c) Total supply decreases, the price increases, welfare decreases.
   d) Total supply increases, the price decreases, welfare increases.
   e) None of the above answers are correct.

12.3.1.5 Exercise 5

Let the demand on an oligopolistic market be given by $P(y) = 300 - \frac{1}{2}y$. The two firms in the market (firm 1 and firm 2) have identical marginal costs of $MC_1 = MC_2 = 30$ and fixed costs of zero. The two firms choose the quantity they supply (Cournot competition).

1. Determine the reaction function of firm 1, $Y_1(y_2)$.
   a) $Y_1(y_2) = \begin{cases} \frac{540-3y_2}{2} & \text{for } y_2 < 400, \\ 0 & \text{else.} \end{cases}$
   b) $Y_1(y_2) = \begin{cases} \frac{540-y_2}{2} & \text{for } y_2 < 540, \\ 0 & \text{else.} \end{cases}$
   c) $Y_1(y_2) = \begin{cases} \frac{1{,}080-y_2}{2} & \text{for } y_2 < 1{,}080, \\ 0 & \text{else.} \end{cases}$
   d) $Y_1(y_2) = \begin{cases} \frac{980-y_2}{2} & \text{for } y_2 < 980, \\ 10 & \text{else.} \end{cases}$
   e) None of the above answers are correct.
2. Determine the Nash equilibrium, in particular, the equilibrium price (p^{CN}) and the equilibrium quantities (y_1^*, y_2^*).
   a) $p^{CN} = 326\frac{2}{3}$ and $y_1^* = y_2^* = 336\frac{2}{3}$.
   b) $p^{CN} = 120$ and $y_1^* = y_2^* = 180$.
   c) $p^{CN} = 165$ and $y_1^* = y_2^* = 135$.
   d) $p^{CN} = 30$ and $y_1^* = y_2^* = 270$.
   e) None of the above answers are correct.

3. Determine the consumer surplus in the Nash equilibrium $CS(y^{CN})$?
   a) The consumer surplus in the Nash equilibrium is indeterminate.
   b) $CS(y^{CN}) = 32{,}400$.
   c) $CS(y^{CN}) = 18{,}000$.
   d) $CS(y^{CN}) = 0$.
   e) None of the above answers are correct.
4. Determine the deadweight loss (DWL) compared to the Pareto-efficient allocation.
   a) $DWL = 8{,}100$.
   b) $DWL = 0$.
   c) $DWL = 60{,}025$.
   d) $DWL = 24{,}000$.
   e) None of the above answers are correct.

12.3.1.6 Exercise 6

Assume the inverse demand function in a Cournot duopoly is given by $P(y) = 100 - y$. Each firm has a cost function of $C(y_i) = 40\, y_i$.

1. Determine firm 1's reaction function. Firm 1's reaction function is:
   a)
$$Y_1(y_2) = \begin{cases} \frac{60-y_2}{4} & \text{for } y_2 < 40, \\ 0 & \text{for } y_2 \geq 40. \end{cases}$$
   b)
$$Y_1(y_2) = \begin{cases} \frac{30-y_2}{2} & \text{for } y_2 < 60, \\ 10 & \text{for } y_2 \geq 60. \end{cases}$$
   c)
$$Y_1(y_2) = \begin{cases} \frac{30-y_2}{3} & \text{for } y_2 < 30, \\ 0 & \text{for } y_2 \geq 30. \end{cases}$$
   d)
$$Y_1(y_2) = \begin{cases} \frac{60-y_2}{2} & \text{for } y_2 < 60, \\ 0 & \text{for } y_2 \geq 60. \end{cases}$$
   e) None of the above answers are correct.
2. Determine the equilibrium price (p^{CN}) and the equilibrium quantity $(y^{CN} = y_1^* + y_2^*)$.
   a) $p^{CN} = 50$, $y_1^* + y_2^* = 50$.
   b) $p^{CN} = 40$, $y_1^* + y_2^* = 60$.
   c) $p^{CN} = 60$, $y_1^* + y_2^* = 40$.
   d) $p^{CN} = 70$, $y_1^* + y_2^* = 30$.
   e) None of the above answers are correct.

3. Both firms want to increase their profit in the market. Thus, they are thinking about behaving collusively. How large could the maximum total profit of both firms be?
   a) $\pi = 450$.
   b) $\pi = 900$.
   c) $\pi = 1{,}000$.
   d) $\pi = 100$.
   e) None of above answers are correct.
4. Before the two firms can collude, a third firm, with an identical cost function, enters the market. Determine the market price (p^{CN}) and the quantity supplied ($y^{CN} = y_1^* + y_2^* + y_3^*$) in the new Nash equilibrium.
   a) $p^{CN} = 60$, $y_1^* + y_2^* + y_3^* = 40$.
   b) $p^{CN} = 50$, $y_1^* + y_2^* + y_3^* = 50$.
   c) $p^{CN} = 40$, $y_1^* + y_2^* + y_3^* = 60$.
   d) $p^{CN} = 55$, $y_1^* + y_2^* + y_3^* = 45$.
   e) None of the above answers are correct.

12.3.2 Solutions

12.3.2.1 Solutions to Exercise 1

- Question 1, answer b) is correct.
- Question 2, answer b) is correct.
- Question 3, answer b) is correct.
- Question 4, answer d) is correct.

12.3.2.2 Solutions to Exercise 2

- Question 1, answer b) is correct.
- Question 2, answer a) is correct.

12.3.2.3 Solutions to Exercise 3

- Question 1, answer b) is correct.
- Question 2, answer d) is correct.
- Question 3, answer c) is correct.
- Question 4, answer b) is correct.
- Question 5, answer b) is correct.

12.3.2.4 Solutions to Exercise 4

- Question 1, answer a) is correct.
- Question 2, answer c) is correct.
- Question 3, answer d) is correct.

12.3.2.5 Solutions to Exercise 5

- Question 1, answer b) is correct.
- Question 2, answer b) is correct.
- Question 3, answer b) is correct.
- Question 4, answer a) is correct.

12.3.2.6 Solutions to Exercise 6

- Question 1, answer d) is correct.
- Question 2, answer c) is correct.
- Question 3, answer b) is correct.
- Question 4, answer d) is correct.

Elasticities

13

This chapter refers to Chapter 14.3 in the textbook.

13.1 True or False

13.1.1 Statements

13.1.1.1 Block 1

1. Elasticities are always independent of the unit of measurement.
2. The point elasticity and the arc elasticity of a linear demand function are always identical.
3. The absolute value of the price elasticity of a demand function $x(p) = \frac{10}{p}$ always equals 10.
4. The price elasticity of a supply function is influenced by the production functions of the firms supplying in the market.

13.1.1.2 Block 2

Market research has measured the following market demand function: $x(p) = 1{,}000 - 300\,p$. The market supply function is $y(p) = \alpha + 100\,p$, with $-\frac{1{,}000}{3} < \alpha < 1{,}000$.

1. The price elasticity of demand is constant.
2. The market demand in equilibrium reacts inelastically to changes in the price.
3. The market supply reacts inelastically to changes in the price if $\alpha > 0$.
4. The market supply in equilibrium reacts elastically to changes in the price if $\alpha < 0$.

M. Kolmar, M. Hoffmann, *Workbook for Principles of Microeconomics*,
Springer Texts in Business and Economics, https://doi.org/10.1007/978-3-319-62662-8_13

13.1.1.3 Block 3

Thilo wants to consistently spend half of his income on apparel.

1. The income elasticity of his demand for apparel is (in absolute terms) 0.5.
2. The price elasticity of his demand for apparel is 0.

Carl wants to spend a constant percentage of his income on apparel.

3. The income elasticity of his demand for apparel is (in absolute terms) 0.5.
4. The price elasticity of his demand for apparel is 0.

13.1.1.4 Block 4

1. $y(p) = p^2$ is a supply function. The supply reacts elastically to price changes.
2. $y(p) = 1 + p$ is a supply function. The supply reacts inelastically to price changes.
3. $x(p, b) = \frac{b}{2p}$ is a demand function. The income elasticity is always equal to 1.
4. $x(p) = 100 - p$ is a demand function. The price elasticity is zero at $p = 100$.

13.1.1.5 Block 5

1. For a linear demand function, the absolute value of the price elasticity is smaller the higher the price is.
2. The price elasticity of the market supply function is independent of the variable costs.
3. If the income elasticity of the demand of a good is negative, then it is a normal good.
4. If the cross-price elasticities of two goods are both negative, then the two goods are substitutes for each other.

13.1.1.6 Block 6

1. If a good is inferior, the income elasticity of demand is larger than zero.
2. If the price elasticity of demand is constant, then it is a linear demand curve.
3. When calculating the price elasticity of demand for a linear demand function, it is irrelevant whether one chooses arc or the point elasticity to calculate its value.
4. A strictly convex demand function of an ordinary good reacts isoelastically to price changes.

13.1.2 Solutions

13.1.2.1 Sample Solutions for Block 1

1. **True**. This true by definition. An elasticity is defined as

$$\frac{\%\text{ change of the dependent variable}}{\%\text{ change of the independent variable}}$$

   See Chapter 14.3.

2. **True**. Let us take a closer look at the definition of point and arc elasticities:
   - Arc elasticity:
$$\epsilon_p^x = \frac{\frac{\Delta x}{x}}{\frac{\Delta p}{p}} = \frac{\Delta x}{\Delta p}\frac{p}{x}$$
   - Point elasticity:
$$\epsilon_p^x = \frac{dx(p)}{dp}\frac{p}{x(p)}$$

   Consequently, there is a difference between point and arc elasticity if the following holds:
$$\frac{\Delta x}{\Delta p} \neq \lim_{\Delta p \to 0}\frac{\Delta x}{\Delta p} = \frac{dx(p)}{dp}.$$
   This always holds if the demand function is non-linear. Once we have a linear demand function, the difference no longer occurs. Hence, the values of arc and point elasticities are one and the same.
   See Chapter 14.3.
3. **False**. The price elasticity of demand is defined as the ratio of a relative demand change to a relative price change (see Definition 14.1 in Chapter 14.3):
$$\epsilon_p^x = \frac{dx(p)}{dp}\frac{p}{x(p)}.$$
   Thus, we get:
$$\begin{aligned}|\epsilon_p^x| &= \frac{10}{p^2}\frac{p}{10/p}\\ &= \frac{10}{p^2}\frac{p^2}{10}\\ &= 1.\end{aligned}$$
4. **True**. The market supply function is influenced by the supplying firms' production function. Thus, the same is true for the price elasticity of the market supply function.

13.1.2.2 Sample Solutions for Block 2

1. **False**.
$$\begin{aligned}\epsilon_p^x &= \frac{dx(p)}{dp}\frac{p}{x(p)}\\ &= -\frac{300p}{1{,}000 - 300p}.\end{aligned}$$
   The price elasticity of demand, ϵ_p^x, depends on the price p and is therefore not constant.

2. **False**. In equilibrium, it is true that

$$\begin{aligned} y(p) &= x(p) \\ \Leftrightarrow \alpha + 100\,p &= 1{,}000 - 300\,p \\ \Leftrightarrow p^* &= 2.5 - \frac{\alpha}{400}. \end{aligned}$$

Regarding the price elasticity of demand, it follws that

$$\begin{aligned} \left|\epsilon^x_{p^*}\right| &= \frac{1}{\frac{1{,}000}{300\,p^*} - 1} \\ &= \frac{750 - \frac{3}{4}\alpha}{250 + \frac{3}{4}\alpha}. \end{aligned}$$

Then,

$$\left|\epsilon^x_{p^*}\right| < 1 \Leftrightarrow \alpha > \frac{1{,}000}{3}.$$

Therefore, the statement is only true if $\alpha > \frac{1{,}000}{3}$. In general, the statement is wrong.

3. **True**. See Definition 14.8 in Chapter 14.3.

$$\begin{aligned} \epsilon^y_p &= \frac{dy(p)}{dp}\frac{p}{y(p)} \\ &= \frac{100p}{\alpha + 100p} < 1 \text{ for } \alpha > 0. \end{aligned}$$

4. **False**. Since $-\frac{1{,}000}{3} < \alpha < 1{,}000$, it must be true that $p^* > 0$ and $x^* = y^* > 0$. We already know from Block 2, Statement 3, that for $p > 0$ $\epsilon^y_p < 1$ if $\alpha > 0$ and $\epsilon^y_p > 1$ if $\alpha < 0$. Hence, this statement must be false.

13.1.2.3 Sample Solutions for Block 3

Both individuals' demand for clothes is $x = \alpha\frac{b}{p}$ (which is derived from $p \cdot x = \alpha b$), in which Thilo's $\alpha = \frac{1}{2}$ and Carl's $\alpha \in [0, 1]$.

1. **False**.

$$\epsilon^x_b = \frac{\partial x}{\partial b}\frac{b}{x} = \frac{1}{2p} \cdot \frac{b}{b/2p} = 1.$$

2. **False**.

$$\epsilon^x_p = \frac{\partial x}{\partial p}\frac{p}{x} = -\frac{b}{2p^2} \cdot \frac{p}{b/2p} = -1.$$

3. **False**.

$$\epsilon^x_b = \frac{\partial x}{\partial b}\frac{b}{x} = \frac{\alpha}{p} \cdot \frac{b}{\alpha b/p} = 1.$$

4. **False.**

$$\epsilon_p^x = \frac{\partial x}{\partial p}\frac{p}{x} = -\frac{\alpha b}{p^2} \cdot \frac{\alpha p}{\alpha b/p} = -1.$$

13.1.2.4 Sample Solutions for Block 4

1. **True.**

$$\epsilon_p^y = \frac{d\,y(p)}{d\,p}\frac{p}{y(p)} = 2\,p \cdot \frac{p}{p^2} = 2.$$

2. **True.**

$$\epsilon_p^y = \frac{d\,y(p)}{d\,p}\frac{p}{y(p)} = 1 \cdot \frac{p}{1+p} = \frac{p}{1+p} < 1, \text{ because } p > 0.$$

3. **True.**

$$\epsilon_b^x = \frac{dx(p)}{db}\frac{b}{x(p)} = \frac{1}{2p} \cdot \frac{b}{\frac{b}{2p}} = \frac{b}{b} = 1.$$

4. **False.**

$$\epsilon_p^x = \frac{dx(p)}{dp}\frac{p}{x(p)} = -1 \cdot \frac{p}{100-p} \quad \text{Hence, } \lim_{p \to 100} \epsilon_p^x = -\infty.$$

13.1.2.5 Sample Solutions for Block 5

1. **False.** For a linear demand function, the absolute value of the price elasticity is smaller the lower the price is. The decisive factors are that (i) $dx(p)/dp$ is constant, because it is a linear demand function and that (ii) the quotient in $p/x(p)$ increases as p increases:

$$\left|\epsilon_p^x\right| = \underbrace{\left|\frac{dx(p)}{dp}\right|}_{=\text{const.}} \cdot \underbrace{\frac{p}{x(p)}}_{\text{increases in } p}$$

2. **False.** The supplying firms' supply functions are influenced by their production functions (and thus the cost functions). Thus, the same is true for the price elasticity of the market supply function.
3. **False.** The income elasticity is, at a given budget b, given as $\epsilon_b^x = \frac{\partial x(b)}{\partial b} \cdot \frac{b}{x(b)}$. If ϵ_b^x is negative, then the same must be true for $\frac{\partial x(b)}{\partial b}$, because $\frac{b}{x(b)} \geq 0$. In this case, however, it is an inferior good. See Definition 4.4 in Chapter 4.2 and Definition 14.3 in Chapter 14.3.
4. **False.** If the cross-price elasticity is negative, then the two goods are complements. Let

$$\epsilon_{p_j}^{x_i} = \frac{dx_i(p)}{dp_j}\frac{p_j}{x_i(p)}, \text{ with } i, j \in \{1, 2\} \text{ and } i \neq j.$$

From $\epsilon_{p_j}^{x_i} < 0$ it follows that

$$\Rightarrow \frac{dx_i(p)}{dp_j} < 0, \text{ since } \frac{p_j}{x_i(p)} \geq 0.$$

If $\frac{dx_i(p)}{dp_j} < 0$, then the goods are complements to each other. See Definition 4.6 in Chapter 4.2 and Definition 14.2 in Chapter 14.3.

13.1.2.6 Sample Solutions for Block 6

1. **False**. If a good is inferior, then the income elasticity is smaller than zero. See Chapter 4.2 and Chapter 14.3.
2. **False**. Linear demand functions do not have a constant price elasticity of demand. While $dx(p)/dp$ is constant for a linear function, $p/x(p)$ is not,

$$\epsilon_p^x = \underbrace{\frac{dx(p)}{dp}}_{=\text{const.}} \underbrace{\frac{p}{x(p)}}_{\neq \text{const.}}.$$

   See Chapter 14.3.
3. **True**. The point and arc elasticities are identical at any point given a linear demand function. See answer to Block 1, Statement 2.
4. **False**. Not every strictly convex demand function is isoelastic.
   Let $x(p) = \frac{1}{p} - 1$. Then,

$$x''(p) = -\left(\frac{1}{p^2}\right)' = \frac{2}{p^3} > 0$$

   and

$$\epsilon_p^x = -\frac{1}{p^2}\frac{p^2}{1-p} = -\frac{1}{1-p}.$$

   Thus, the demand function is not isoelastic.

13.2 Open Questions

13.2.1 Problems

13.2.1.1 Exercise 1

The demand for the concert tickets of a specific pop singer can be divided into two groups: enthusiasts (E) and conformists (C). The quantities of tickets demanded by these two groups are displayed in Table 13.1.

Table 13.1 Exercise 1. Demand for concert tickets

Price	Demand E	Demand C
100	1,000	3,750
90	1,500	4,250
80	2,000	5,000
70	2,500	6,000
60	3,000	7,200
50	3,500	9,000
40	4,000	12,000

1. Draw a graph to visualize the demand function of each group.
2. Calculate the price elasticity of demand for both groups for the following variations of the price for concert tickets:
   a) an increase from 80 to 100 Swiss Francs,
   b) an increase from 60 to 70 Swiss Francs,
   c) an increase from 40 to 50 Swiss Francs.
   Explain your results.

13.2.1.2 Exercise 2

Suppose one has the following demand functions for two individuals:

$$x^A(p) = \alpha\, p^{-\beta}, \qquad x^B(p) = a - b\, p,$$

with $\alpha, \beta, a, b > 0$.

1. Calculate the price elasticity of demand for infinitesimal changes in the market price (point price elasticity of demand). Illustrate the elastic and the inelastic parts of demand in a graph.
2. Let $\alpha = 5{,}000$, $\beta = 1$, $a = 6{,}000$, and $b = 50$.
   a) Calculate the arc price elasticity of demand for a discrete (non-infinitesimal) change of the market price. Starting from a price of 100 Swiss Francs, suppose that a price change of 10 Swiss Francs occurs.
   b) Compare your results in Question 2a) with the corresponding point price elasticity. Comment on your answer.

13.2.1.3 Exercise 3

Christian has a monthly budget of b, which he always uses completely to buy pizza (P), wine (W) and economics textbooks (E).

1. Pizza and wine are complementary goods. What does this say about the cross-price elasticities of Christian's demand for these goods?
2. All goods are normal goods and Christian's demand for pizza and wine reacts elastically to income changes. What does this tell you about the income elasticity of Christian's demand for economics textbooks?
3. Is it possible for his income elasticity of demand to be negative for all goods at once?
4. What relation do you see between the income and price elasticity of demand, on the one hand, and the type of a good, on the other hand?

13.2.1.4 Exercise 4

Calculate the price and income elasticities (and the cross-price elasticity, where possible) for the following demand functions (p_1 and p_2 are the prices and b refers to the income):

1. $x_1(p_1, b) = \frac{b}{2\,p_1}$, with $p_1 > 0, b > 0$,
2. $x_1(p_1, p_2, b) = \frac{b^2}{p_1+p_2}$, with $p_1 > 0$, $p_2 > 0, b > 0$,
3. $x_1(p_1, b) = b - p_1$, with $b > p_1 > 0$.

13.2.1.5 Exercise 5

Suppose there are two countries (A and B) with the following demand functions for the same good:

$$x^A(p) = 100 - p \text{ and } x^B(p) = 160 - 2\,p.$$

The supply functions are identical in both countries and are given as

$$y(p) = -20 + p.$$

1. Calculate the market-clearing equilibrium prices (p^{A^*} and p^{B^*}) and the quantities traded in equilibrium ($x^A(p^{A^*}) = x^{A^*}$ and $x^B(p^{B^*}) = x^{B^*}$) in both countries. Draw a graph to illustrate the market equilibrium.
2. Calculate the price elasticity of demand ($\epsilon_p^{x^A}$ and $\epsilon_p^{x^B}$) and supply (ϵ_p^y) for both countries in equilibrium.
3. The governments in both countries levy a tax of $t = 5$ Swiss Francs per unit, to be paid to the government by the producers. Thus, the price that suppliers request (gross price, i.e., the market price including the tax) rises by 5 Swiss Francs. The new supply curve is now:

$$y_t(p) = -20 + p - t = -25 + p.$$

   a) Calculate the new market-clearing equilibrium price ($p_t^{A^*}$ and $p_t^{B^*}$) and the quantities traded in equilibrium ($x^A(p_t^{A^*}) = x_t^{A^*}$ and $x^B(p_t^{B^*}) = x_t^{B^*}$) in both countries. Calculate the market-clearing price net of tax as well, i.e., the price that is received by the producers ($q_t^{A^*} = p_t^{A^*} - t$ and $q_t^{B^*} = p_t^{B^*} - t$). Draw a graph to illustrate your results.
   b) What is the correlation between the price elasticity of demand ($\epsilon_{p^{A*}}^{x^A}$ and $\epsilon_{p^{B*}}^{x^B}$), on the one side, and the decline in the quantity of goods purchased ($x^{A^*} - x_t^{A^*}$ and $x^{B^*} - x_t^{B^*}$), on the other side?
   c) What is the correlation between the difference in the old and new gross prices ($p_t^{A^*} - p^{A^*}$ and $p_t^{B^*} - p^{B^*}$), on the one side, and the relation of the price elasticity of demand and supply ($\epsilon_{p^{A*}}^{x^A}, \epsilon_{p^{A*}}^{y}$ and $\epsilon_{p^{B*}}^{x^B}, \epsilon_{p^{B*}}^{y}$), on the other side?

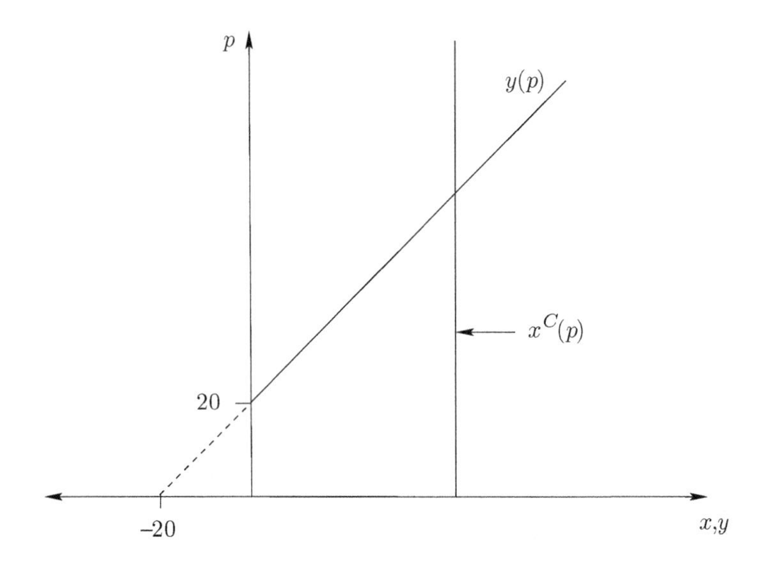

Figure 13.1 Exercise 5.4. Country C

4. Figures 13.1 and 13.2 display markets in countries C and D. Assume that the supply function corresponds to those in countries A and B.
   a) What is the price elasticity of demand in countries C and D?
   b) Suppose that a unit tax $t = 5$ is levied. What are the effects on prices and the quantity of goods traded in equilibrium? Illustrate your answers in the graphs.
   c) Do you see your hypothesis from Questions 3b) and 3c) disproved?

13.2.2 Solutions

13.2.2.1 Solutions to Exercise 1

1. Price-quantity diagram: Fig. 13.3.
2. The price elasticity of demand, for a specific, non-marginal change of the market price, can be calculated via the arc elasticity:

$$\epsilon_p^x = \frac{\frac{\Delta x}{x}}{\frac{\Delta p}{p}},$$

   with x and p representing the values of quantity and price *before* the price change.

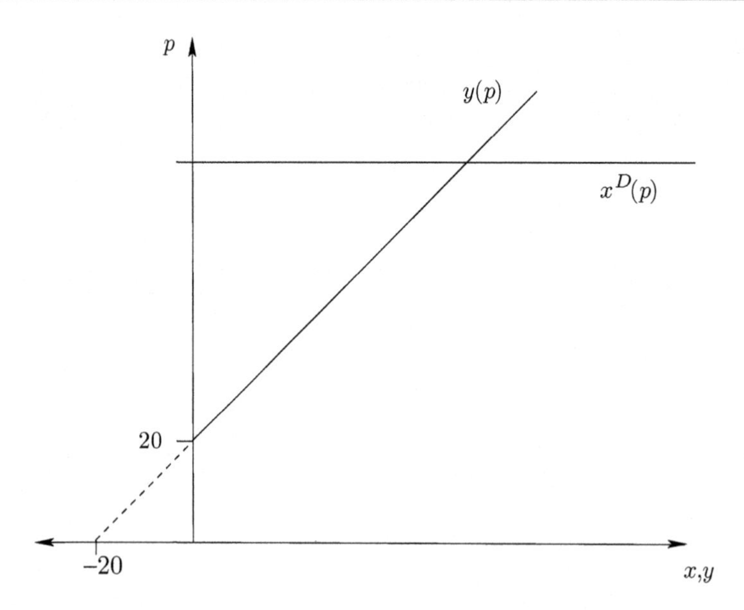

Figure 13.2 Exercise 5.4. Country *D*

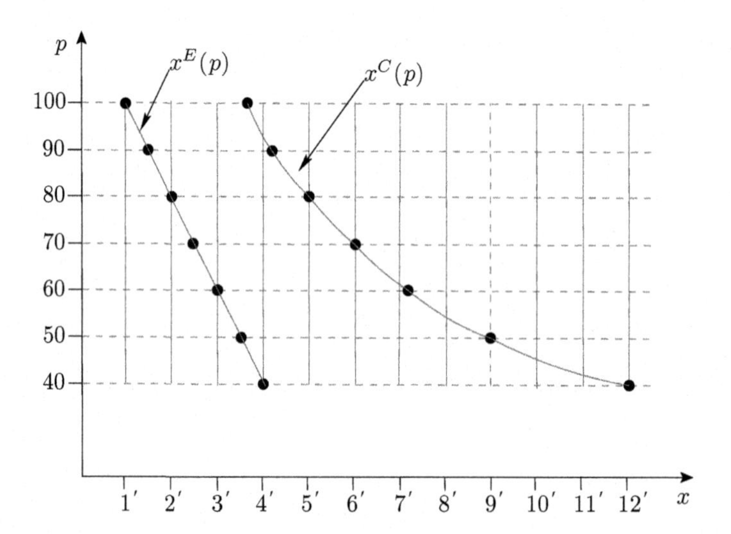

Figure 13.3 Exercise 1.1. Price-quantity diagram

We thus get the following for the E-group:

a)

$$\epsilon_p^{x^E} = \frac{\frac{-1{,}000}{2{,}000}}{\frac{20}{80}} = \frac{-\frac{1}{2}}{\frac{1}{4}} = -2,$$

b)

$$\epsilon_p^{x^E} = \frac{\frac{-500}{3{,}000}}{\frac{10}{60}} = \frac{-\frac{1}{6}}{\frac{1}{6}} = -1,$$

c)

$$\epsilon_p^{x^E} = \frac{\frac{-500}{4{,}000}}{\frac{10}{40}} = \frac{-\frac{1}{8}}{\frac{1}{4}} = -0.5.$$

For the C-group, we find that:

a)

$$\epsilon_p^{x^C} = \frac{\frac{-1{,}250}{5{,}000}}{\frac{20}{80}} = \frac{-\frac{1}{4}}{\frac{1}{4}} = -1,$$

b)

$$\epsilon_p^{x^C} = \frac{\frac{-1{,}200}{7{,}200}}{\frac{10}{60}} = \frac{-\frac{1}{6}}{\frac{1}{6}} = -1,$$

c)

$$\epsilon_p^{x^C} = \frac{\frac{-3{,}000}{12{,}000}}{\frac{10}{40}} = \frac{-\frac{1}{4}}{\frac{1}{4}} = -1.$$

Clearly, the demand of the E-group reacts inelastically, as well as elastically, towards price changes. That means that, contingent on the pre-change values of x and p, the percentage change in x is either larger (given a price increase from 80 to 100 Swiss Francs), equal (given a price change from 60 to 70 Swiss Francs) or smaller (given a price increase from 40 to 50 Swiss Francs) than the corresponding percentage price change.

The demand in the C-group is isoelastic: the percentage change in x always equals the percentage price change.

13.2.2.2 Solutions to Exercise 2

1. The definition of the price elasticity of demand is:

$$\epsilon_p^x = \frac{d\,x(p)}{d\,p}\frac{p}{x(p)}.$$

Thus, we get the following for the individuals A and B:

$$\epsilon_p^{x^A} = -\alpha\,\beta\,p^{-\beta-1}\,p\,\alpha^{-1}\,p^{\beta} = -\beta.$$

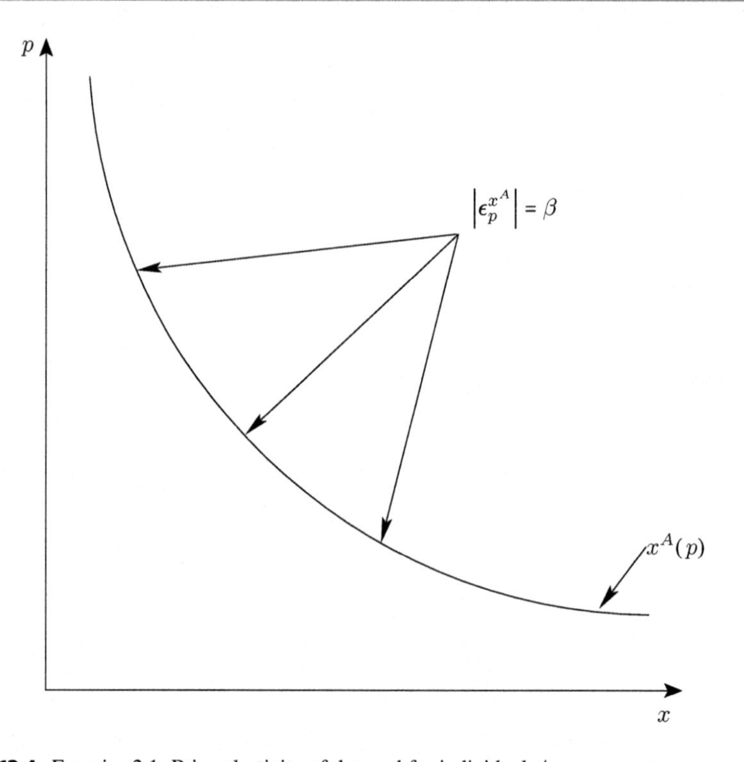

Figure 13.4 Exercise 2.1. Price elasticity of demand for individual A

and

$$\epsilon_p^{x^B} = -b\,\frac{p}{a - b\,p} = \frac{1}{1 - \frac{a}{b\,p}}. \tag{13.1}$$

Contingent on the value of β, the demand of A either always reacts inelastically to price changes, or always elastically (see Fig. 13.4). However, the demand of B reacts elastically as well as inelastically, contingent on the pre-change value of x and p (see Fig. 13.5). From Eq. 13.1, it immediately follows that:

$$\left|\epsilon_p^{x^B}\right| \begin{cases} > 1 & \text{if } p > \frac{a}{2b}, \\ = 1 & \text{if } p = \frac{a}{2b}, \\ < 1 & \text{if } p < \frac{a}{2b}. \end{cases}$$

Hence, the demand of B reacts elastically (inelastically) to price changes if $p > \frac{a}{2b}$ $(p < \frac{a}{2b})$.

2. Inserting the values of the parameters, we get:

$$x^A(p) = \frac{5{,}000}{p} \text{ and } x^B(p) = 6{,}000 - 50\,p.$$

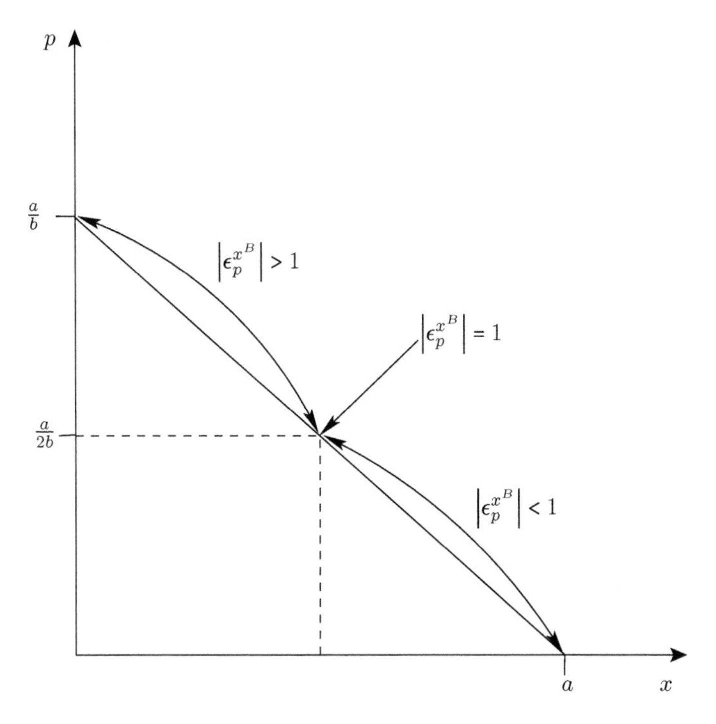

Figure 13.5 Exercise 2.1. Price elasticity of demand for individual B

a) • $x^A(100) = 50$, $x^A(110) = 45\frac{5}{11}$, so that $\Delta x = -4\frac{6}{11}$. We get:

$$\epsilon_p^{x^A} = \frac{\frac{\Delta x}{x}}{\frac{\Delta p}{p}} = \frac{\frac{-4\frac{6}{11}}{50}}{\frac{10}{100}} = -\frac{10}{11} \approx -0.91.$$

• $x^B(100) = 1{,}000$, $x^B(110) = 500$, so that $\Delta x = -500$. We get:

$$\epsilon_p^{x^B} = \frac{\frac{\Delta x}{x}}{\frac{\Delta p}{p}} = \frac{\frac{-500}{1{,}000}}{\frac{10}{100}} = -5.$$

b) From Question 1, one can conclude that:

$$\epsilon_p^{x^A} = -\beta = -1,$$

and

$$\epsilon_p^{x^B} = \frac{1}{1 - \frac{a}{b\,p}} = \frac{1}{1 - \frac{6{,}000}{50 \cdot 100}} = -5.$$

For individual A, both measures of elasticity yield different values. The opposite holds for individual B. Why is that? Let us take a closer look at the definitions of point and arc elasticity:

- Arc elasticity: $\epsilon_p^x = \dfrac{\frac{\Delta x}{x}}{\frac{\Delta p}{p}} = \dfrac{\Delta x}{\Delta p}\dfrac{p}{x}$
- Point elasticity: $\epsilon_p^x = \dfrac{dx}{dp}\dfrac{p}{x}$

Consequently, there is a difference between point and arc elasticity if the following holds:

$$\frac{\Delta x}{\Delta p} \neq \lim_{\Delta p \to 0} \frac{\Delta x}{\Delta p} = \frac{dx(p)}{dp}.$$

This always holds if the demand function is non-linear.

13.2.2.3 Solutions to Exercise 3

1. If two goods are complements, the demand for one good falls as the price of the other good increases.

$$\frac{\partial x_P(p_P, p_W, b)}{\partial p_W} < 0 \qquad \text{and} \qquad \frac{\partial x_W(p_P, p_W, b)}{\partial p_P} < 0.$$

The cross-price elasticities are then given as:

$$\epsilon_{p_W}^{x_P} = \underbrace{\frac{\partial x_P(p_P, p_W, b)}{\partial p_W}}_{<0} \underbrace{\frac{p_W}{x_P(p_P, p_W, b)}}_{\geq 0} \leq 0,$$

and

$$\epsilon_{p_P}^{x_W} = \underbrace{\frac{\partial x_W(p_P, p_W, b)}{\partial p_P}}_{<0} \underbrace{\frac{p_P}{x_W(p_P, p_W, b)}}_{\geq 0} \leq 0,$$

whereas the second term is never negative, because price and quantity are always non-negative. Thus, the cross-price elasticity is either zero or negative.

2. If goods are normal, demand increases in income:

$$\frac{\partial x_P(p_P, p_W, b)}{\partial b} > 0 \qquad \text{and} \qquad \frac{\partial x_W(p_P, p_W, b)}{\partial b} > 0.$$

The income elasticity of, for example, pizza is:

$$\epsilon_b^{x_P} = \underbrace{\frac{\partial x_P(p_P, p_W, b)}{\partial b}}_{>0} \underbrace{\frac{b}{x_P(p_P, p_W, b)}}_{\geq 0} \geq 0.$$

We assume that $\epsilon_b^{x_P} > 1$ and $\epsilon_b^{x_W} > 1$.
Let us take a look at Chris's budget:

$$b = x_P p_P + x_W p_W + x_E p_E. \tag{13.2}$$

The left-hand side (LHS) of Eq. 13.2 represents Chris's budget, while the right-hand side (RHS) represents Chris's expenditures. If the LHS increases by one percent, so does the RHS. However, this is only possible if demand reacts inelastically to the income changes.

3. When all goods have a negative income elasticity of demand, total expenditures decrease when income increases. However, since Christian always spends his entire budget, this can never happen and at least one good has to have a positive income elasticity of demand.
4. In general, there are two goods (1 and 2) and income (b).
   - The income elasticity of demand for good 1 is:

$$\epsilon_b^{x_1} = \frac{\partial x_1(p_1, p_2, b)}{\partial b} \frac{b}{x_1(p_1, p_2, b)}.$$

   We also know that a good is normal (inferior) if demand reacts positively (negatively) to an increase in income. Thus, we know that:

$$\frac{\partial x_1(p_1, p_2, b)}{\partial b} \frac{b}{x_1(p_1, p_2, b)} \left\{ \begin{matrix} > \\ < \end{matrix} \right\} 0 \quad \Rightarrow \quad \frac{\partial x_1(p_1, p_2, b)}{\partial b} \left\{ \begin{matrix} > \\ < \end{matrix} \right\} 0,$$

   since $\frac{b}{x_1(p_1, p_2, b)} \geq 0$. Hence, the good is normal (inferior) if the income elasticity of demand is positive (negative).
   - The cross-price elasticity of demand for good 1 is defined as:

$$\epsilon_{p_2}^{x_1} = \frac{\partial x_1(p_1, p_2, b)}{\partial p_2} \frac{p_2}{x_1(p_1, p_2, b)}.$$

   We also know that goods 1 and 2 are substitutes (complements) if demand for good 1 (2) reacts positively (negatively) to an increase in the price of good 2 (1). Thus, we know that:

$$\frac{\partial x_1(p_1, p_2, b)}{\partial p_2} \frac{p_2}{x_1(p_1, p_2, b)} \left\{ \begin{matrix} > \\ < \end{matrix} \right\} 0 \quad \Rightarrow \quad \frac{\partial x_1(p_1, p_2, b)}{\partial p_2} \left\{ \begin{matrix} > \\ < \end{matrix} \right\} 0,$$

   since $\frac{p_2}{x_1(p_1, p_2, b)} \geq 0$.
   Hence, goods 1 and 2 are mutual substitutes (complements) if the cross-price elasticity of the demand for good 1 (2) is positive (negative).

- The price elasticity of demand is defined as:

$$\epsilon_{p_1}^{x_1} = \frac{\partial x_1(p_1, p_2, b)}{\partial p_1} \frac{p_1}{x_1(p_1, p_2, b)}.$$

We also know that good 1 and good 2 are ordinary (Giffen goods) if their demand decreases (increases) as their price increases. Thus, we know that:

$$\frac{\partial x_1(p_1, p_2, b)}{\partial p_1} \frac{p_1}{x_1(p_1, p_2, b)} \left\{ \begin{matrix} > \\ < \end{matrix} \right\} 0 \quad \Rightarrow \quad \frac{\partial x_1(p_1, p_2, b)}{\partial p_1} \left\{ \begin{matrix} > \\ < \end{matrix} \right\} 0,$$

since $\frac{p_1}{x_1(p_1,p_2,b)} \geq 0$. Hence, good 1 and good 2 are ordinary (Giffen goods) if their own-price elasticity of demand is negative (positive).

13.2.2.4 Solutions to Exercise 4

1. The elasticities are:
   - Price elasticity:

$$\epsilon_{p_1}^{x_1} = \frac{\partial x_1(p_1, b)}{\partial p_1} \frac{p_1}{x_1(p_1, b)} = -\frac{b}{2\,p_1^2} \frac{2\,p_1^2}{b} = -1,$$

   - Income elasticity:

$$\epsilon_{b}^{x_1} = \frac{\partial x_1(p_1, b)}{\partial b} \frac{b}{x_1(p_1, b)} = \frac{1}{2\,p_1} \frac{2\,p_1\,b}{b} = 1.$$

2. The elasticities are:
   - Price elasticity:

$$\begin{aligned} \epsilon_{p_1}^{x_1} &= \frac{\partial x_1(p_1, p_2, b)}{\partial p_1} \frac{p_1}{x_1(p_1, p_2, b)} = \frac{-b^2}{(p_1 + p_2)^2} \frac{p_1(p_1 + p_2)}{b^2} \\ &= -\frac{p_1}{p_1 + p_2}, \end{aligned}$$

   - Cross-price elasticity:

$$\begin{aligned} \epsilon_{p_2}^{x_1} &= \frac{\partial x_1(p_1, P_2, b)}{\partial p_2} \frac{p_2}{x_1(p_1, p_2, b)} = \frac{-b^2}{(p_1 + p_2)^2} \frac{p_2(p_1 + p_2)}{b^2} \\ &= -\frac{p_2}{p_1 + p_2}, \end{aligned}$$

   - Income elasticity:

$$\epsilon_{b}^{x_1} = \frac{\partial x_1(p_1, p_2, b)}{\partial b} \frac{b}{x_1(p_1, p_2, b)} = \frac{2\,b}{p_1 + p_2} \frac{b(p_1 + p_2)}{b^2} = 2.$$

3. The elasticities are:
   - Price elasticity:

$$\epsilon_{p_1}^{x_1} = \frac{\partial x_1(p_1, b)}{\partial p_1} \frac{p_1}{x_1(p_1, b)} = -\frac{p_1}{b - p_1} = \frac{p_1}{p_1 - b},$$

   - Income elasticity:

$$\epsilon_{b}^{x_1} = \frac{\partial x_1(p_1, b)}{\partial b} \frac{b}{x_1(p_1, b)} = \frac{b}{b - p_1}.$$

13.2.2.5 Solutions to Exercise 5

1. - Country A:

$$\begin{aligned} & x^A(p) = y(p) \\ \Leftrightarrow \quad & 100 - p = -20 + p \\ \Leftrightarrow \quad & p = 60. \end{aligned}$$

   Thus, the equilibrium price is $p^{A^*} = 60$. The equilibrium value of x is then:

$$x^{A^*} = x^A(p^{A^*}) = 100 - 60 = 40.$$

   This is illustrated in Fig. 13.6.
   - Country B:

$$\begin{aligned} & x^B(p) = y(p) \\ \Leftrightarrow \quad & 160 - 2\,p = -20 + p \\ \Leftrightarrow \quad & p = 60. \end{aligned}$$

   The equilibrium value of x is:

$$x^{B^*} = x^B(p^{B^*}) = 160 - 120 = 40.$$

   This is illustrated in Fig. 13.7.
2. The price elasticities are:
   - $\epsilon_{p^A}^{x^A} = \dfrac{-p}{100 - p}$. In equilibrium we get

$$\epsilon_{p^{A*}}^{x^A} = \frac{-p^{A^*}}{100 - p^{A^*}} = -\frac{3}{2}.$$

   - $\epsilon_{p^B}^{x^B} = \dfrac{-2\,p}{160 - 2\,p}$. In equilibrium we get:

$$\epsilon_{p^{B*}}^{x^B} = \frac{-2\,p^{B^*}}{160 - 2\,p^{B^*}} = -3.$$

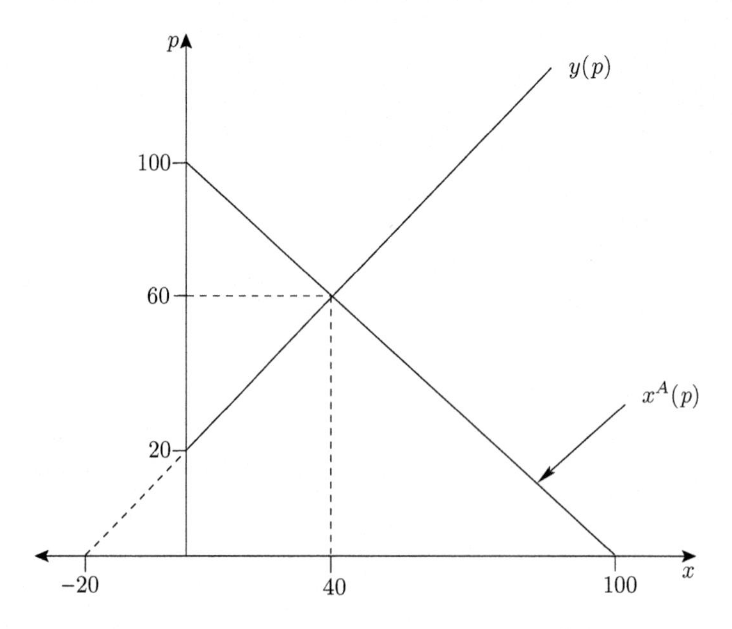

Figure 13.6 Exercise 5.1. Market equilibrium in country *A*

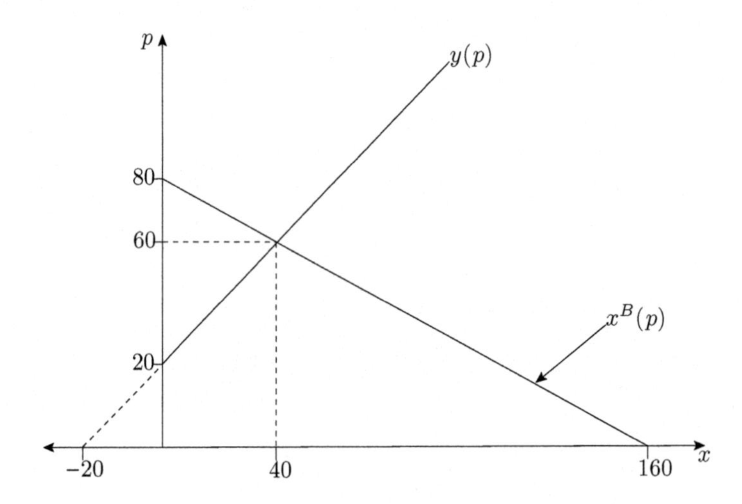

Figure 13.7 Exercise 5.1. Market equilibrium in country *B*

- $\epsilon_p^y = \dfrac{p}{p-20}$. In equilibrium we get:

$$\epsilon_{p^{A*}}^y = \frac{p^{A^*}}{p^{A^*}-20} = \frac{3}{2} \qquad \text{and} \qquad \epsilon_{p^{B*}}^y = \frac{p^{B^*}}{p^{B^*}-20} = \frac{3}{2},$$

since $p^{A^*} = p^{B^*}$.

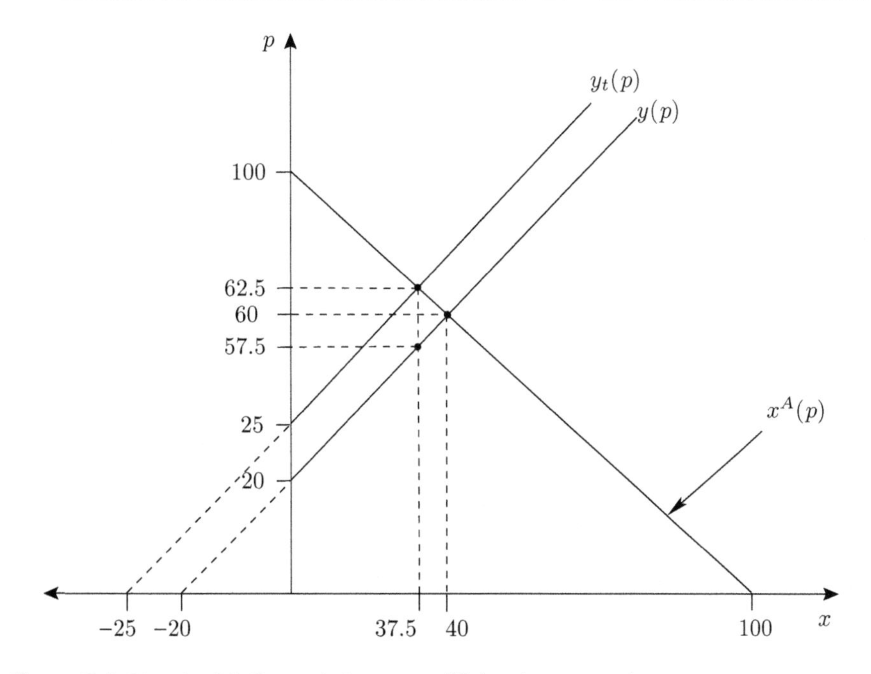

Figure 13.8 Exercise 5.3. Pre- and after-tax equilibrium in country A

3. After tax equilibrium:
   a) • Country A: The gross price is given as:

$$\begin{aligned} & x^A(p) = y_t(p) \\ \Leftrightarrow \quad & 100 - p = -25 + p \\ \Leftrightarrow \quad & p = 62.5. \end{aligned}$$

Hence, the equilibrium after-tax gross price is $p_t^{A^*} = 62.5$. The after-tax net price in equilibrium (that is the price producers receive in the end) is then:

$$q_t^{A^*} = p_t^{A^*} - t = 62.5 - 5 = 57.5.$$

The equilibrium value of x is:

$$x_t^{A^*} = x^A(p_t^{A^*}) = 100 - 62.5 = 37.5. \tag{13.3}$$

This is illustrated in Fig. 13.8.

   • Country B:

$$\begin{aligned} & x^B(p) = y_t(p) \\ \Leftrightarrow \quad & 160 - 2\,p = -25 + p \\ \Leftrightarrow \quad & p = 61\frac{2}{3}. \end{aligned}$$

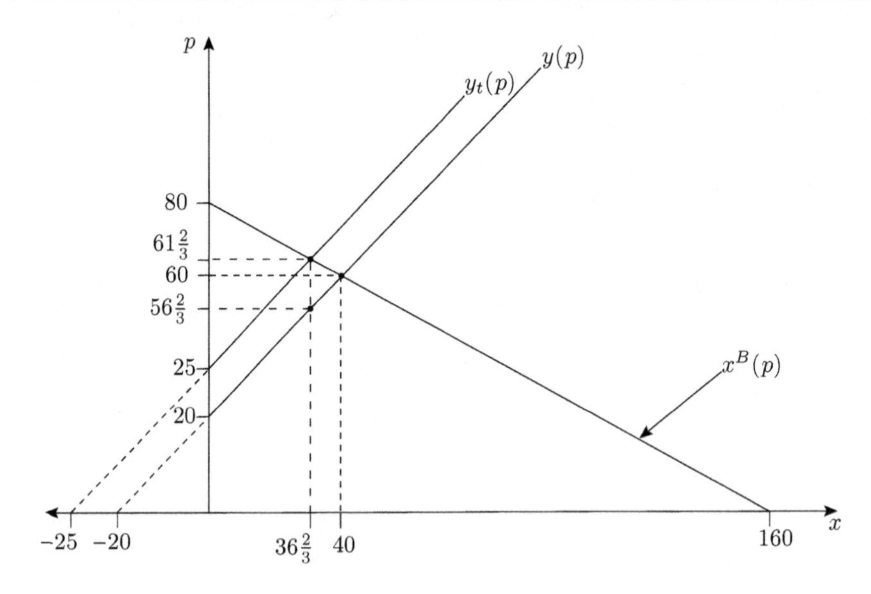

Figure 13.9 Exercise 5.3. Pre- and after-tax equilibrium in country B

The gross price is thus $p_t^{B^*} = 61\frac{2}{3}$. The net price in equilibrium is:

$$q_t^{B^*} = p_t^{B^*} - t = 61\frac{2}{3} - 5 = 56\frac{2}{3},$$

and

$$x_t^{B^*} = x^B(p_t^{B^*}) = 160 - 123\frac{1}{3} = 36\frac{2}{3}.$$

This is illustrated in Fig. 13.9.

b) The demand in country B reacts more elastically than the demand in country A does, because:

$$\left|\epsilon_{p^{B*}}^{x^B}\right| = 3 > \frac{3}{2} = \left|\epsilon_{p^{A*}}^{x^A}\right|.$$

The decline of goods purchased in country B is larger than the decline in country A, because:

$$x^{B^*} - x_t^{B^*} = 3\frac{1}{3} > 2.5 = x^{A^*} - x_t^{A^*}.$$

Thus, we can formulate the following hypothesis: *the more elastic demand reacts to price changes in the pre-tax equilibrium, the larger the reduction in the goods purchased will be.*[1]

[1] Of course, this only holds if we assume that the price elasticity of supply in equilibrium ($\epsilon_p^y(p^{A^*})$ and $\epsilon_p^y(p^{B^*})$) is identical in all countries.

c) In country A, the price elasticity of demand and supply are identical:

$$\left|\epsilon_{p^{A*}}^{x^A}\right| = \epsilon_{p^{A*}}^{y} = \frac{3}{2}.$$

The after-tax gross price in equilibrium ($p_t^{A^*}$) equals the pre-tax price plus half of the unit tax. Thus, consumers and producers share the burden of the tax equally. Consumers have to pay 2.5 Swiss Francs more than before, while producers receive 2.5 Swiss Francs less than before.
In country B, demand reacts more elastically than supply, because:

$$\left|\epsilon_{p^{B*}}^{x^B}\right| = 3 > \frac{3}{2} = \epsilon_{p^{B*}}^{y}.$$

The after-tax gross price in equilibrium ($p_t^{B^*}$) equals the pre-tax equilibrium price (p^{B^*}) plus $1\frac{2}{3}$ Swiss Francs. However, this is only $\frac{1}{3}$ of the unit tax. Thus, the consumers' tax burden is lower than that of the producers: while consumers have to pay $1\frac{2}{3}$ Swiss Francs more than before the tax, producers receive $3\frac{1}{3}$ Swiss Francs less than before the tax.
Thus, we can formulate the following hypothesis: *the more elastic demand reacts to price changes compared to the supply, the less pronounced the increase in the gross price will be and the stronger the reduction in the net price will be.*

4. For countries C and D, we find that the following holds:
   a) Demand in country C (D) reacts completely inelastically (elastically) to an increase in the market price:

$$\epsilon_p^{x^C} \to 0 \qquad \text{and} \qquad \epsilon_p^{x^D} \to -\infty.$$

   b) Again, the unit tax is $t = 5$, which has to be paid to the government by the producers.
      - Country C. Since the demand in country C reacts completely inelastically, producers shift the tax burden towards the consumers:
        - The gross price equals the pre-tax equilibrium price plus the unit tax ($p_t^{C^*} = p^{C^*} + t$),
        - the net price remains constant ($q_t^{C^*} = p^{C^*}$),
        - and the equilibrium value of x remains constant ($x_t^{C^*} = x^{C^*}$).

        These results are illustrated in Fig. 13.10.
      - Country D. Since demand in country D reacts completely elastically, producers cannot shift the tax burden towards the consumers:
        - The gross price stays constant ($p_t^{D^*} = p^{D^*}$),
        - the net price equals the old equilibrium price reduced by the value of the tax ($q_t^{D^*} = p^{D^*} - t$),
        - and the value of x decreases ($x_t^{D^*} < x^{D^*}$).

        These results are illustrated in Fig. 13.11.

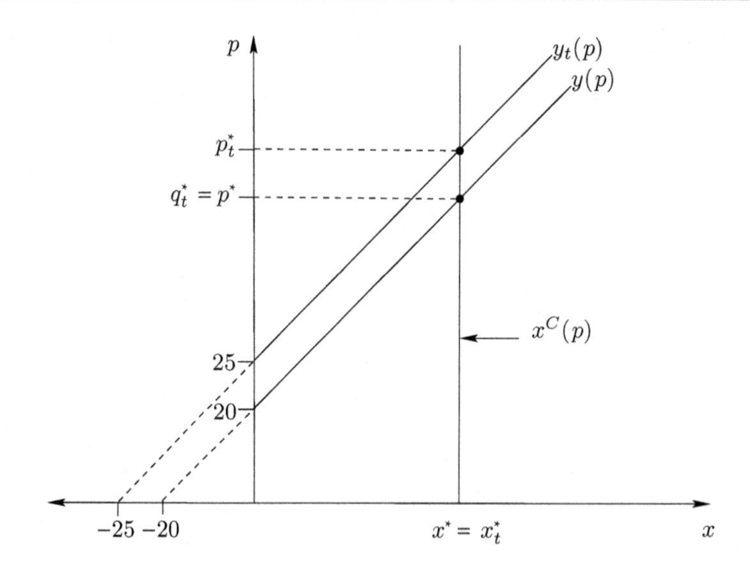

Figure 13.10 Exercise 5.4b). Pre- and after-tax equilibrium in country C

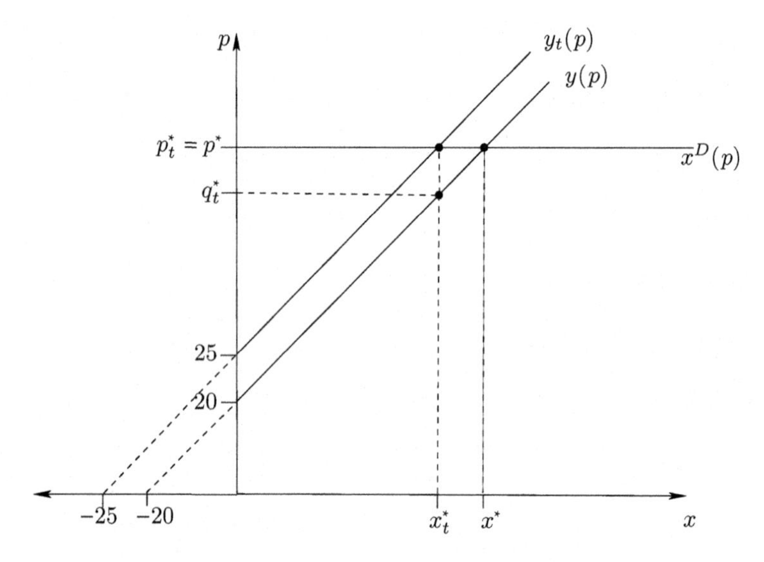

Figure 13.11 Exercise 5.4b). Pre- and after-tax equilibrium in country D

c) The hypothesis derived in Question 3b) was: *the more elastic demand reacts to price changes in the pre-tax equilibrium, the larger the reduction in the goods purchased will be.* This still holds in countries C and D. The hypothesis derived in Question 3c) was: *the more elastic demand reacts to price changes compared to the supply, the less pronounced the increase in the gross price will be and the stronger the reduction in the net price will be.* Again, this still holds in countries C and D.

Printed by Printforce, the Netherlands